Islands

Isl

ands

A

TREASURY

OF

CONTEMPORARY

TRAVEL

WRITING

CAPRA PRESS
SANTA BARBARA

Printed in the United States of America.

Book design and typography by Albert Chiang
Cover design and illustration by Deja Hsu.
Typesetting by Stanton Publication Services.

LIBRARY OF CONGRESS CATALOGING–IN–PUBLICATION DATA

Islands: a treasury of contemporary travel writing / editors of
ISLANDS Magazine
p. cm.
ISBN 0–88496–349–7 : $12.95
1. Islands. I. Theroux, Paul.
G500. I86 1992
910' .9142--dc20 91–38011
CIP

CAPRA PRESS
P. O. Box 2068, Santa Barbara, CA. 93120

CONTENTS

Caribbean

Europe

Asia/Africa

Foreword

BY THE EDITORS OF ISLANDS MAGAZINE

IMAGINE YOURSELF ON A DESERT ISLAND. A BIT OF WHITE SAND, A PALM TREE or two, expanses of blue-green water all around, and not a soul in sight. A simple image, almost a cliché. But what a powerful image! Strong enough to launch, in the autumn of 1981, a magazine called ISLANDS. Since then that image has summoned up an eager corps of adventurous writers and photographers, and nourished the dreams of thousands of loyal readers.

Over the last decade the image has broadened. With more than half a million islands to choose from, ISLANDS has crisscrossed the globe, looking not only for the idyllic but also the unexpected. The craggy rock outcrops of the North Atlantic have appeared as counterpoint to the azure waters of the Caribbean. A city island in Europe, rich in culture and history, makes a remote South Pacific atoll—where few people have ever set foot—seem all the more fantastic.

The stories in this collection reach back to the magazine's very first issue, which included Herbert Gold's irrepressible exploration of the Seychelles, the scattered Indian Ocean archipelago that for many of us represents the epitome of a dream escape. And the stories illustrate the magazine's global reach, covering islands off every continent in the world.

More important, with their blend of literary style and personal passion, they represent the best of contemporary travel writing—the perfect match of place and writer. Thus, Nantucket becomes for Jan Morris the very soul of islandness, while the remote islands of the South Pacific turn into a microcosm of the world for Paul Theroux. Frances FitzGerald laces her acute political insights with a love of the underwater world when she visits Vanuatu, and Pico Iyer's lyrical prose provides the ideal prism through which to view the lively contradictions of Cuba.

The writers remind us that islands are more than mere landscape. Our experiences of them are human experiences, recounted with humor—as in Christopher Buckley's journey along Australia's Great Barrier Reef, and Lawrence Millman's stay on the Cook Islands—or retold with an island's characteristic accent. Never has the music in Newfoundland speech been played back with such affection as in Carol McCabe's tale. And if the Irish love of words and word play is common knowledge, Jessica Maxwell renews the idea with witty exhilaration.

In the end, it barely matters whether we will ever get to these places. Adam Nicolson's loving contemplation of his own Shiant Islands, the Scottish islands he inherited, makes that brilliantly clear. His dots of land may be virtually unvisitable, but they have an unwavering hold on his imagination—and ours. Let there be no mistake. Islands are a source of dreams. They inspire.

—THE EDITORS

Note: The year of original publication appears at the end of each article.

Pacific

A World of Islands in the South Pacific

PAUL THEROUX

THE IDEAL ISLAND IS A whole world, and what a world. Size is incidental. Where insularity is concerned, completeness is everything, and even a tiny island may contain multitudes. Never mind Manhattan. There are wilder places that are prettier, and much more habitable, and a great deal odder.

Every island is different. Islands may be small, but they have strange configurations and are often found in very awkward places. Look at Easter Island. Look at Tristan da Cunha. (But don't be misled by the islet of Langerhans; it is found in the pancreas.) Many islands are volcanoes, and some are furiously active. This does not deter the islanders. Just last year lava poured down the main street of Kalapana, on Hawaii's Big Island, and wiped out the whole town, but no one seriously blamed the volcano. It was felt fatalistically that this was all the doing of Pele, the violent and lovely goddess of fire. "I believe that Madame Pele must keep creating," one man said after his house had been destroyed, echoing the thoughts of other residents.

In order to get the widest possible experience of such singularity, I traveled to the Pacific Ocean, which is a world of islands, and I took my collapsible kayak as part of my luggage. My intention was to paddle around and see these islands from every angle.

When there was a closely clustered archipelago I paddled from one island to another; and when a tiny offshore island glittered in the sun I paddled out to it. In many parts of the Pacific—American Samoa is a prime example—not many people paddle or sail canoes these days. So my arrival through the surf was often a grand entrance, and then everyone would want to try paddling my boat.

A kayak can handle most sea conditions. High wind is a much greater deterrent than surf. "What about sharks?" people constantly asked me. I was never bothered by sharks, though I saw them from time to time, notably in Fiji and New Guinea. And I saw slithering sea snakes and spiders and large lizards.

Pacific islanders rarely talk about sharks or saltwater crocodiles, or any other biting creatures. The worst menace on any Pacific island is the mosquito, and malaria the most dangerous disease. That's why I traveled with a tent. And I actually preferred it, and my mosquito netting, to sleeping in a village hut or a bad hotel. I grew to like camping on the beach, falling asleep to the sound of waves.

"What is this?" I was asked by numerous customs inspectors when they saw all my canvas bags.

"A boat," I always replied. This usually raised a smile.

At Fua'amotu airport in Tonga I was charged "wharfage" because I was carrying this vessel. Wharfage in this case (one 15-foot kayak) was two dollars.

In the Solomon Islands I paddled out from Guadalcanal to Savo, a small mound in the sea that is an active bubbling and steaming volcano surrounded by palm-leaf huts. Every morning hundreds of moorhen-like birds, called megapodes (big feet), dig deep holes in the sandy beach at the edge of the volcano and lay an egg in each one.

The birds take no further interest in the egg, but the sand is geothermally hot, just enough to incubate the egg. Yet few ever hatch: After the birds fly into the trees, the people on this island flock to the beach and dig holes three or four feet down and unearth the eggs, which are even bigger than duck eggs. Some they sell, some they trade, and some they carry higher to the hot springs of the volcano to hard-boil and eat.

"I'm going down to the egg fields," the men say in the morning. Egg digging is their chief occupation. It is very unlucky (the men believe) for Savo women to dig for megapode eggs. But that is just a male confidence trick. The egg fields are full of guys together, laughing, fooling like Phi Gams or the Elks Club brothers.

"We have everything," the people say. They don't want roads or cars or an airstrip. They have no use for telephones. They don't want electricity; they don't want television, though they have a yen for videos.

They have fruit trees, nut trees, fertile soil, and fish. They have instant hot water and all the eggs they can eat. In some villages they worship the megapode bird and sacrifice pigs to it on jungle altars. ("This place is very taboo. Look, but don't go close," a villager told me, as he brushed a four-inch spider from his cheek.) Their eggs are also a comfortable income, each one worth one Solomon Islands dollar.

The people have few yearnings. "What would you do if you had a spare 50 dollars?" I asked a man on Savo.

His reply was instant: "I would go to Guadalcanal and watch a *Rambo* video."

The people are mostly Melanesian, almost African looking (an Ethiopian stowaway lived on Guadalcanal for two years as a Solomon Islander before anyone got wise to him), and if you ask them where they came from they won't point to the horizon (as Polynesians do) and say their ancestors sailed from the distant *moana*—the ocean—in double-hulled canoes. They will say: We came from a shark, or we came from a bird, or a snake.

Sharks are summoned to the harbor by the wise men in some Solomon Islands villages. The sharks are fed: Pigs are slaughtered and thrown to them, and the sharks are appeased. In the villages where snakes are venerated, a strange ritual takes place each time someone takes a long journey. He or she goes to the snake hole and thrusts a hand into the darkness and takes hold of the fattest snake and mumbles a prayer. If you don't do that, something terrible might happen to you on your journey. But since these snakes are very venomous, something terrible could also happen to you before you set sail. In other Solomon Islands villages people hurry to church and sing themselves hoarse.

"I am free on an island," a Tanna man told me, "and I can control my own life."

This was in Vanuatu, formerly the New Hebrides, in dusky, friendly Melanesia, where I was paddling my little boat. Tanna was one of the oddest places I have ever been, which is saying a great deal. It is proof that even on the smallest island all sorts of differing life-styles can flourish simultaneously. Never mind the political parties (there were three on Tanna), or the one Peace Corps volunteer, or Breffni McGeough, the kava-drinking Irishman who runs one of the local hotels. There were much odder things on Tanna.

The volcano on Tanna belches fire, brimstone, and sulfurous fumes. The volcano is called Yasur. ("Yasur" means God.) Not far from it are villages where the women wear grass skirts and nothing else, the men only a strategic bunch of grass that looks like a whisk broom. Big Numbers, they are called, pronounced "nambas." Elsewhere there are people called Small Numbers.

They have no interest in any religion but their own, which concerns the sun and the moon and the harvest of roots and bananas (and involves stamping dances and occasional sacrifices of animals). Now and then earnest missionaries bring them Bibles and try to teach them Christian prayers.

"What do you do when missionaries visit?" I asked the chief.

"We listen to them," he said, "and, when they leave, we pray in our own way." He puffed on a banana-leaf cigar and said, "We don't mind if they are Christians."

But there are militant non-Christians on this island. One night in the 1930s, the stories go, a little man stepped out of the bush near Green Point on Tanna. His name was Jon Frum. He approached an islander, a Presbyterian.

"Throw away your Bible," Jon Frum said. "Don't listen to the foreign missionaries. Return to your old ways."

Jon Frum was apparently an American. He said he would bring "goods" to the people of Tanna if they returned to their own ways. A so-called cargo cult sprang up around him. Villages dedicated themselves to Jon Frum; they put up a big cross, painted it red—this was Jon Frum's symbolic color—and sang songs about him. During the war Americans came to these islands, and they seemed to prove the truth of Jon Frum's promises: Americans were bringing lots of goods to the island!

The Jon Frum villages on Tanna began flying American flags. Even now on Fridays and Saturdays islanders stay up all night singing, urging Jon Frum to come back.

THERE WERE ALL SORTS OF OTHER PEOPLE ON TANNA. ONE DAY I WAS HEADED out to one of the muddy villages. It was raining hard, the road was bad, but the distance was not too far. I came across a broken-down van. Four smiling New Zealand men stood near it. I said hello.

"Hello," they said, and one added, "Bless you."

"What's the problem?"

"No problem," the bearded leader said.

The van was stuck. The rain was slashing into us. The men were soaked to the skin. No problem!

"He's trying to make obstacles for us," the bearded man said.

"Who is?"

"The devil." And the man smiled at the rain. "We are Christians, from Auckland. The devil knows we're going to hold a service up there in north Tanna. But he won't succeed with his obstacles."

And the others laughed confidently as the rain pelted the jungle around us. In earlier days these people might have ended up in a pot—many did.

The people on Tanna were once cannibals. "Some people here are still cannibals, but they won't tell you," Chief Tom Namake told me. His best story was about a cannibalistic group of Tanna villagers who killed an Englishman and then instead of devouring him became obsessed with the idea of cooking and eating his canvas shoes, which they felt were a great delicacy.

Chief Tom wore a T-shirt with a quotation from Isaiah on it ("He made me into a polished arrow"), but said he believed in magic most of all. He personally felt that the people on Tanna had originated from two magic sticks, long ago.

"Some people say we came from Africa," he said. "How could that be so? We get seasick going from island to island!"

Just east of Tanna there was a small island—no more than a rock in the ocean. It was called Futuna. The few hundred people were Polynesian, their language quite similar to that of people 2,000 miles away. I asked a man from Futuna where his ancestors came from, and he said promptly, "Tonga, and before that, Asia."

Futuna was in the middle of nowhere, but it was also part of the great Polynesian community of peoples who ranged from Easter Island to Hawaii, and through Tahiti to the Maoris of New Zealand. Polynesian culture is particular: It is language, it is food, and it is religion, which seems to take root on an island in a way it seldom does on a continent. Cut off from the challenges of doubt and discussion, it continues unchanged.

Polynesians practice a muscular Christianity, the same sanctimonious fundamentalism that they were taught a century and a half ago by fervent Victorian missionaries. Travel writers—and travel agents—may continue to popularize the Pacific as a place of great sensuality, but my impression of those islands is of churches and cholesterol, of Christianity and canned corned beef, of tubby evangelists Bible-thumping to an almost deafening degree.

It would be unthinkable for a woman to go bare breasted in Tonga, but it is actually unlawful for a man not to wear a shirt. What about bathing suits? Well, Tongans tend to swim wearing all their clothes, and it is not unusual to see a woman jump into the sea wearing a skirt, and a blouse with a T-shirt underneath.

I found the Tahitians among the most straitlaced folk in the Pacific. (The expatriates are an altogether different story.)

The Samoans are rather fussy and fastidious about their Polynesian customs, but they are even fussier about their Christianity. Even a small Samoan village may have several churches, and each one will have two or three services on Sunday.

It was also in Tonga and Samoa that various valuable items were stolen from me: my Walkman, my western belt with a silver buckle, some money, some camping equipment. I complained to a German expat: "And they're supposed to be so religious!" "If you steal a lot, you have to pray a lot," he reasoned.

Nothing is more pacific than a Pacific Sunday: No store is open, no work is done. No business may be transacted in Fiji. Quiet must be observed in a Samoan village. Checks in Tonga can't be dated on Sunday, or they would be invalid. And an American I met in Nuku'alofa had been told it would be sacrilegious for him to jog on Sunday. The Tongan airlines timetable is blank for Sunday, nor are any ships unloaded that day. The Cook Islanders are great hymn singers, and all are Christians. The Marquesans have even turned the Catholic Mass into something that resembles a Polynesian folk opera.

Everyone is welcome in Polynesia, though after a period of time the stranger is encouraged to move on—go "off-island," in the Pacific phrase. Even New Zealanders are receiving rather persistent hints of this kind from militant islanders who want to rename New Zealand Aotearoa, "The Land of the Long White Cloud," in Maori—perhaps not a bad idea. The Fijians are trying to serve notice on the people of Indian extraction (third and fourth generation) who live there.

It is rare that land in Polynesia may be sold. The majority of American Samoans live in the United States, but even so they laugh at the idea of any non-Samoan wishing to buy land; it is impossible. In Tonga, in the Cook Islands, in Fiji, the land may be leased but not sold. The land is us, the people say; we are the land.

Island land is sacred. This may be true the world over. Every acre of it is owned and accounted for. It may look wild, but it is not wilderness. There is not much of it, and so it is your wealth, your history, your identity. Ethnic Hawaiians feel this way, but most of their land is gone, sold to those other islanders, the sons of Nippon, who not long ago coveted the whole of the Pacific.

"Go back to your fields," the Solomon Islanders were told by the Japanese solders in 1942. An old man in Guadalcanal related the memory to me. "Grow coconuts, grow vegetables, catch fish. We will buy food from you," the Japanese said.

"What else did the Japanese tell you?" I asked.

"They said," and at this the old man smiled, " 'We are staying here forever.' "

The Battle of Guadalcanal uprooted them, of course, but at the great cost of more than 30,000 lives. Yet there is not a viable island in the entire Pacific that does not have the commercial presence of Japanese, and from Fiji to Hawaii hotel workers are urged to learn a few polite phrases in Japanese.

Even in Pearl Harbor, Hawaiians try to think of ways of being nice to the Japanese. I was there for the 49th anniversary of the bombing last December, and everyone was on good behavior ("Though we sometimes have a problem with Japanese tourists, who don't realize how deeply we feel about this place," a guard told me. "They talk loud and sometimes laugh and fool at the Arizona Memorial.") Deep down the Hawaiians' feelings are quite different—how could they forget or forgive such an outrage?—but few are so crass as to mention it. Courtesy is regarded as a virtue not a weakness in Hawaii, and no one grovels.

Generally speaking, islands breed a unique kind of peacemaking and politeness. Rudeness tends to be a big-city survival technique—a form of assertiveness, or the expedient of people who know they can disappear into the hinterland. But islands have little or no hinterland. People keep running into each other. The person you insult today might be your golf partner tomorrow. For better or worse, islanders are family members and for their own sanity try to practice conflict avoidance. Even in crowded Honolulu it is rare for someone to blow a horn at you, and when it happens it will almost certainly not be a local person. Horn-blowing is regarded as one of the more obnoxious *haole* habits.

Conspicuousness is something I have come to associate with islands. You may go to an island as a way of going into exile, but it is hard to hide on an island, as Paul Gauguin discovered. Gauguin was so disappointed by the bustling bourgeois society of Tahiti he headed for the remote Marquesan island of Hiva Oa, where of course he stuck out like a sore thumb. Anyone who wishes to remain anonymous would be well advised to avoid setting up house on an island. If there is someone you want to meet very badly on an island you only have to stay a while and you will run into him or her eventually.

"YOU MUST SEE THE CABINET MINISTER," I WAS TOLD IN HONIARA, IN THE SOLOMONS, when I said I wanted to camp on a particular island. In any mainland metropolis this would be a near impossibility. It seemed the height of presumption to barge into a cabinet minister's office only to ask him whether I could go camping. On an island it is a simpler matter.

The minister was not in his office, but his secretary obligingly gave me his home phone number. We arranged to meet, we talked half the morning about world affairs, and off I went in my little boat to the island, with my tent, and a case of beer, and a letter of introduction from the minister to his brother, who—it turned out—was illiterate. A bystander read the letter out loud, and the brother welcomed me in pidjin English.

"Who is in charge of this island?" I asked in various places, and I was always led directly to the house of that person. He was the paramount chief on some islands, or the president, or the mayor, but he was nearly always a sitting duck. It used to be said that Marlon Brando lived in obscurity in French Polynesia. That's the opposite of the truth. His house could not be more obvious, and there is a bungalow resort on his island, Tetiaroa. Even after the scandal of her lover's murder by her brother, Cheyenne Brando still listed her telephone number in the Papeete telephone directory. (I called her up but got no reply.)

On the island of Kauai everyone knows Sylvester Stallone's house, and Tom Selleck's outside Honolulu is fairly prominent. People are constantly dropping in to Bengt Danielsson's house in Tahiti to ask him how it was when he sailed there with Thor Heyerdahl on the *Kon-Tiki*.

On the remote atoll of Aitutaki, 150 miles from Rarotonga, in the Cook Islands, I was dragging my kayak up the beach when I saw a tall and somewhat familiar figure, apparently beachcombing. It was a stoutly built European man, and we soon fell into conversation.

It is simply not possible for anyone to snub you on a little atoll in the Pacific.

"You look familiar," I said.

"I used to be prime minister of New Zealand," he replied. "I'm David Lange."

I introduced myself.

"Oh, God, don't tell me you're here writing a book about Aitutaki," he said, and he seemed genuinely alarmed that I might be doing that. I had the impression that he was planning a book of his own.

I said, no, I was just paddling around for my health. This put his mind at rest, and we spent the next three days discussing world affairs, and world leaders, and ("as the author of so many books about trains, you'll love this") what he called "the ultimate railway story."

I can't do his anecdote complete justice, since Mr. Lange is a great mimic and has a gift for storytelling—a brilliant sort of rapid-fire delivery. And the birds were twittering in Aitutaki's coconut palms, as he cheerily told me his macabre tale. The story concerned a trip he had once taken in India as a student, traveling from Delhi to Bombay.

"In those days they had a wonderful dining car, with heavy silver and cloth napkins and waiters running to and fro," he said. He chose the beef curry, violating Cardinal Rule One of traveling in vegetarian India: Never eat meat. (Rule Two: Never drink plain water. Rule Three: Never eat salad. Rule Four: Never eat fruit you haven't cut or peeled yourself.)

"The meat tasted strange, but of course it had been heavily spiced. I was violently ill afterwards," Mr. Lange went on, "and I spent days in bed. I have never been so ill with food poisoning. And I wasn't the only one.

Most of the people who had the beef curry on that train ended up in the hospital."

He chuckled at the memory and said, "About a month later I read that one of the waiters on the Delhi to Bombay run had been arrested for supplying dismembered human corpses to the dining car, claiming they were fresh beef. They were hardly recognizable, of course, after they had been turned into curry."

If we had not been on an island I doubt that I would have met him; and I think it was because we met on an island that we became friends. Something about that isolation made our meeting companionable, and it was possible in a short time for this acquaintance to ripen into friendship. In the end it was David Lange who drove me and my boat to the airport in a borrowed, beat-up Jeep—he was my farewell delegation—and we promised to meet again soon. If not on the North Island of New Zealand, then in Honolulu on the island of Oahu.

In Tonga, the only monarchy in the Pacific, I was eager to meet the king and be granted an audience. I knew him by his great reputation: He was an intellect, a modernizer, known for his geniality as well as for his long silences. He was the eldest son of Tonga's celebrated Queen Salote, who had more or less upstaged Queen Elizabeth at her own coronation in London in 1953. It rained hard that day. Tongan custom insists that in order to show respect you must demonstrate humility, and you cannot imitate the person you are honoring. At the first sign of rain, Queen Elizabeth's footmen put up the hood on her carriage as it rolled toward Westminster Abbey.

Hoods were raised on the rest of the carriages—all but one, that of the Queen of Tonga. She sat, vast and saturated and majestic, in a carriage that was awash; and from that moment she earned the love and affection of every person in England.

I asked about the crown prince of Tonga.

"He's around," a Tongan told me. "You will find him in a restaurant or a club."

"What about the king?"

"If his flag is on the flagpole of the palace"—it is perhaps the only wooden palace in the world—"that means he is inside."

The flag, the royal standard, was fluttering on the flagpole, so I went to the office of the king's private secretary and asked whether I could have an audience. We talked a little bit about what I would ask him, and a Tongan friend vouched for me; and a week later I was being led into the royal compound by the king's aide-de-camp.

I found the king blunt and full of life. He was physically overwhelming, his vast bulk surmounted by an enormous head. He had a pair of glasses in each pocket, a gold wristwatch on each hand, and he wore a Tongan skirt. He peered at me through heavy-lidded eyes, and he often laughed uproariously at a joke he had made. He had a barreling, bullfrog laugh.

"The French are totally unreliable!" he said, and his laughter boomed through the palace. And to prove it he gave me a thumbnail summary of the Franco-Prussian War, as well as the details of the way the French had sunk the Greenpeace ship, *Rainbow Warrior*, in New Zealand.

What about French nuclear testing in the Pacific?

"It is wrong. They have to do it somewhere else."

What about the French colonizing Polynesia?

"They must leave," the king said. "It will be painful financially for the Polynesians, but it is the right thing. The French must go."

"Sooner or later, Your Highness?"

"Soon," the king said.

When I prefaced another question by claiming that he was an absolute monarch, the king denied that this was so. He had a parliament, he said. (He appoints every member, of course.) But a few minutes later he went on to say that the reason he was able to modernize Tonga—television and a new air terminal, perhaps plans for an oil refinery, and oil drilling—was that he had no foot-dragging committees.

"I am the committee," he said. "A committee of one. There are no dissenting votes!" Another booming laugh.

When I left the palace, I reflected that I had never tried to meet a king before—how many kings are there in the world, in any case?—and that I had been successful purely because he was an island king. An island is a kind of fortress, but once you are on it nothing is closed to you, and it seems possible to gain entrance anywhere, to meet anyone at all. That may not be universally true of islands (although I suspect that in the rest of the world the exceptions are few), but it seems to be the rule in the Pacific.

MEETING A KING, HOWEVER, IS NOT THE ULTIMATE ISLAND EXPERIENCE. IN MY opinion, the ultimate is having an island to yourself. I conceived the ambition of paddling offshore and making camp on a deserted island and of living there on my own for a week or so.

I flew with my boat to the Vava'u Group, in the north of the Tongan archipelago—about 50 islands, big and small, many of them uninhabited. I had a good chart of the islands that I had bought in Honolulu, and I had a tent and a sleeping bag and insect repellent. I bought a week's supply of food in the administrative center, Neiafu, and I set off, paddling through the driving rain. It was the hurricane season. I met fishermen in dugouts and outrigger canoes, and they told me which islands were uninhabited. Of course, the islands were owned by someone, and I would have to ask permission to camp, but, as long as I had the courtesy to ask, permission would be granted.

In my mind I wanted an island with cliffs, with a safe sandy beach; an island with jungle, with birds. Ideally, it would have drinking water, but just in case I had my own—gallons of it in a water bag. People sometimes accused me of being naive and a little foolish, alone in a kayak in waters I had never paddled before. But my kayak was more seaworthy and stable when it was loaded with gear than it was when it was empty, and I felt completely independent, completely free, carrying my own household, so to speak, my own water and food; my own means of getting from island to island.

A little piece of land, Pau Island, loomed in the mist as I paddled through Vava'u. I stopped at a village on a nearby island and asked permission to camp there. Yes, the chief said, no problem. I distributed little presents to the villagers and then headed across the gray expanse of water to set up camp on the island of Pau. I liked the name. It was almost "Paul." I pitched my tent out of the wind, on high ground, at the edge of a sandy beach. Nearby were eight-foot—easily climbable—coconut palms, with coconuts perfect for picking.

In the following days I picked green coconuts and cut holes in the husks and drank the sweet water. I cooked over my little kerosene stove. I went snorkeling and bird-watching. My island was the haunt of hundreds of fruit bats, called flying foxes. I watched them squabbling and chattering by day as they hung from the trees; and at nightfall they flew silently to other islands to feed. I visited the uninhabited islands a few miles away. I listened to the Gulf War on my short-wave radio. I crawled into the tent before eight o'clock when the mosquitoes were thickly swarming, and I got up at six. I developed a little island routine of working and napping.

One day a yacht anchored offshore on a sandbar. I paddled out to warn them that as the tide was ebbing they would soon be aground. We introduced ourselves. They knew one of my brothers! They snapped my picture and left. Apart from that single instance of it's-a-small-world, I lived the life of Robinson Crusoe.

And, like Crusoe, I would have been much happier sharing this island with one other person, but circumstances didn't permit it.

So I made the best of it, and enjoyed it, inventing a schedule of things to do, breaking the day into three parts, eating regularly, so I would not go off my head. I liked the island for not having any dogs or roosters.

Afterward, people asked: Weren't you frightened?

"Yes, once," was my reply.

Crusoe was petrified when he saw a footprint on his island and realized that it was not his own. I had a related experience. One day at dusk, after the tide had ebbed all afternoon, I looked up from my meal (couscous and baked beans, with canned mackerel) and saw footprints everywhere. They led down the beach and into the woods; up to the cliffs, along the shore, across the dunes, all around the camp, desperate little solitary tracks. There were hundreds, perhaps thousands of footprints, and what frightened me—what eventually impelled me to break camp soon afterward and head for the nearest inhabited island, where I was assured of a welcome (in the Pacific no stranger is turned away)—what sent a chill through me, was the thought that every single footprint, every urgent little trail, was mine. [1991]

The Cook Islands

LAWRENCE MILLMAN

WITH THE DECOR OF PARAdise, as with cuisine, there is no disputing matters of taste. One person's caviar is another person's soggy dumpling. For myself, I've always preferred a good hearty slog over rugged terrain to the more easeful charms of the beach. Likewise I've never thought the coconut palm—emblem of the paradisiacal tropics—any more interesting, of its kind, than the maidenhair tree, which smells like slightly soiled feet, or the arctic rowan, which after a hundred years is still only eight inches high.

So it was that upon my arrival in Rarotonga, capital island of the Cooks, I went for a backcountry trek rather than a stroll along one of the island's proverbially empty, palm-clad beaches. My guide for this trek was a man named Pa Teuraa. Dressed in *pareu* loincloth, barefoot, and wielding a machete, he looked the very model of an old-style Polynesian. Except for his dreadlocks. He wore his hair this way, he told me, so it'd be easier to wash. Then he pointed to the cluster of red florets that herbally inclined islanders like himself use as a shampoo. He also pointed to the bright red flower, known in Maori as *ta'akura,* that definitely should not be used as a shampoo. It harbors the spirit of a woman by the same name who long ago was raped by some bloody-minded Rarotongan warriors. She died as a result and, legend has it, passed into this flower. If you were to put a ta'akura into your hair, Ta'akura herself would appear and lure you to a death no less brutal than her own, usually, Pa said, by flinging you off a mountain or impaling you on a reef. Needless to say, I didn't put one of these pretty little flowers in my hair.

Rarotonga is a high volcanic island, with razorback ridges and dense jungle, more or less on the Tahitian design. And, like Tahiti, a few parts of its interior still have not been walked by man or woman. Pa and I were not headed into these parts, but where we were headed seemed almost as unfrequented.

We traversed fields of thick bracken, crossed brisk streams, climbed muddy embankments, passed through disused taro plantations and the stony remnants of old villages, until at last we came to our destination—a boulder of undeniable heft and, according to Pa, undeniable religious significance. Here the Ancient Ones had worshiped their gods in premissionary times, probably as long ago as the great Maori voyages to New Zealand. Here, too, the Modern Ones seemed to have left some of their own gods, including empty rum bottles, worn-out sneakers, and a T-shirt that advertised the Pacific Resort, Muri Beach, Rarotonga.

I shook my head and got ready to be appalled. To think, rubbish in such a sacred place! Yet Pa said it wasn't rubbish, but offerings made by his fellow Rarotongans to propitiate the gods. See, even a couple of *ariki* (chiefs) had inscribed their names on the boulder's face.

"We call that graffiti," I said.

"We call it giving a piece of yourself," Pa said.

He, too, left a piece of himself or some sort of offering when he came here, and said I should do the same thing. So I left my socks, since they were covered with impossible-to-remove *piri-piri* petals anyway. And all at once, miraculously, socklessly, I felt much better. As if the gods were letting me know that they were indeed propitiated. (Equally miraculous, I never saw any other rubbish during my visit to the Cook Islands.)

A CURIOUS GROUP OF ISLANDS, BUT JUST HOW CURIOUS? PEOPLE HERE HONOR the Sabbath even more virtuously than Scottish Highlanders, but they also honor Tangaroa, ancient god of fertility, whose well-endowed figure appears on their one-dollar coin.

Their House of Parliament is no bigger than a convenience store, and their government patrol boat is named, happily, *Te Kukupa* ("the dove"). They change their names to commemorate events: the death of a sibling, a finger nipped off by a coconut crab, or perhaps ownership of a motor scooter with a faulty clutch. They have a positive horror of rubbish, even a lone candy bar wrapper in the front yard, and yet they also inter deceased family members in this same front yard, a place they consider more intimate than a distant cemetery plot.

THE COOK ARCHIPELAGO IS STREWN IN ALMOST PERVERSELY LANGUOROUS FASHion across a swath of southern Polynesia, strewn as if by a Creator more attuned to the sounds of a coconut-shell ukelele, all sweetness and lassitude, than to the demands of His own work ethic. From the air the Cooks seem to go on forever, volcanic islands, flat islands, atolls, coral *motus*, isles of copra, and isles of pearl shell, no two alike. It took Western navigators some 240 years to track them all down. Even today there persists a rumor of one more as-yet-undiscovered Cook, a blessedly heathen atoll somewhere in the emptiness of the Pacific. Blessedly, too, the old beachcomber who told me about this atoll refused to tell me where it was, so that both of us could nurture our fantasies of lost worlds.

"The most detached parts of the earth," Capt. James Cook called these islands, though he named them the Herveys, after a lord of the admiralty, rather than after himself. And detached they remain 200 years later, not only from New Zealand, which governs their external affairs (they look after their own internal business), but also from each other. The northernmost Cook, Penrhyn lies almost 1,000 miles north of the southernmost one, Mangaia. Mangaia itself is Rarotonga's closest neighbor, but at 150 miles it's not what you'd call real neighborly.

Such distances make the 14 inhabited islands of the Cooks feel like a crazy quilt of small nations, each with its own traits and prejudices, instead of a generic country. We all know that Frenchmen are supposed to be excitable, Swiss industrious, and Swedes dour. Well, Pukapukans are supposed to be sexually exuberant and Mangaians sexually backward. Aitutakians are voluble, even glib, and love to argue themselves into untenable positions so they can argue themselves out again. Rarotongans are haughty, because it was from Ngatangiia, on their island, that the first great exodus of Maori canoes to New Zealand set forth. Maukeans, on the other hand, are docile and even a little stupefied, as befits a people with the unenviable reputation of being the only South Sea islanders ever known to have dived for taro roots.

And what of the inhabitants of lonely Palmerston (pop. 66)? Ever since 1862, they've been Marsterses. Before that date, the atoll supported crabs, frigate birds, and bêches-de-mer, but no people. Then came an Englishman named William Marsters with his two Polynesian wives. He acquired a third Polynesian wife from a Portuguese seaman, whereupon he resolved to populate the atoll through his own initiative. The old fellow succeeded quite admirably in his endeavor. The present-day inhabitants of Palmerston, including the Reverend Bill, are almost all Marsterses, and the atoll may be the very best place in the world to go if you want to hear the speech of 19th-century Lancashire, England.

Concerning Atiuans, they were supposed to be unimaginably fierce, at least before the apostle of the Pacific, the Reverend John Williams, calmed them down with Christianity. They would just as soon kill and eat a stranger as look at him. One of their favorite pastimes was to raid docile Mauke and force Maukean wives to eat their own husbands, though the selfish Atiuans kept the best cuts of husband for themselves. Even now the most common expression in the Cooks for a fight, a ruckus, or just a quarrel is: "*E Atiu oki.*" ("It's Atiu trouble.")

But on Atiu I didn't encounter any trouble at all. No one gazed at me with either homicidal or gastronomic intent. Quite the contrary. I could hardly go anywhere without being offered a lift on someone's motor scooter; even huge boisterous mamas draped unsparingly over their vehicles would come to a rattling halt and invite me on. Nor could I go anywhere without a rich variety of foodstuffs being forced into my arms. The day I walked Atiu's version of the Appian Way, a coastal track overhung with ironwood trees and purple sprays of bougainvillea, I returned to my digs laden with breadfruit, taro, yams, and utos (coconuts germinated in the shell)—gifts from locals who figured that I must be a singularly deprived sort of traveler to be venturing out with just a camera.

Only Atiu's *makatea* keeps up the island's reputation for fierceness. One day I walked this makatea with a guide named Tangi Jimmy. Actually Tangi Jimmy walked; I inched along delicately. Makatea consists of coral thrust up onto land eons ago and gradually fossilized, now into pinnacles and battlements, now into gargoyles no less fantastic than Notre Dame's. Parts of it are so sharp that if you fall down, you'll carry a souvenir of Atiu on your person for a long, long time. Other parts contain small grottolike forests and even limestone caves. Tangi Jimmy and I were making our way toward one of these caves, Atiu's largest, called Anatakitaki.

It may seem odd that in the sunny South Pacific, I should be opting for the dark underground. In my defense I can only say that the sun is universal, the same blatant star whether it's shining in the Cook Islands or Cook County, Illinois. Anatakitaki, however, is the only place in all the universe where I'd ever see a kopeka, a species of swift nearly as rare as the whooping crane. No more than 200 kopekas live in this deep, multichambered cave, navigating their passage through it with echo sounders, like bats. Once outside, they turn off these echo sounders and navigate more conventionally, though (tireless birds!) they don't land until they're back home in the gloom.

Flashlights in hand, we entered this gloom ourselves and almost immediately began scrambling over boulders. Long palm roots, like strings, hung in our eyes. Stalactites also hung in our eyes. Because of these stalactites and the slippery-damp boulders, we had to move with a certain caution, looking up and down simultaneously, doing what cavers call the "Groucho walk." After half an hour or so of this posture, we began seeing kopekas—small brown blurs hovering along the cave's calloused ceiling. They were sounding their echo devices repeatedly. *Click click click click click.*

Soon we came to a large chamber so full of clicking that it sounded as if an orchestra of castanet players had taken up residence there. We switched off our flashlights and sat listening to this rather unusual concert. Tangi Jimmy told me about the time he'd gotten this far in the cave when all of a sudden his light gave out. On his hands and knees he had crawled back to the entrance by following the clicks of kopekas who were leaving the chamber themselves and heading outside for an insect dinner.

ON MY LAST EVENING ON ATIU, I VISITED ANOTHER LOCAL RARITY, ALBEIT A RARity less avian than cultural—a *tumunu*. Though once a cherished institution throughout the Cooks, nowadays tumunus survive only in Atiu. The word refers to the pandanus-thatch huts where men gather to discuss, say, whether or not to put a new road through the taro swamp. It also refers to the home brew that fuels these discussions. Finally, it refers to the hollowed-out palm tree stumps in which this home brew, usually made from oranges, is left to ferment. Tumunus are supposedly illegal, but on the evening of my visit the island constable put in an appearance and threw back a few coconut cups of spirits himself.

The coconut cup went back and forth, back and forth, between the *tangata kapu* (barman) and his increasingly less sober patrons. At one point the tangatu kapu asked all of us to bow our heads, and then he thanked the Almighty for giving us this home brew even as he asked His forgiveness because we were drinking it. At another point all eyes focused on me, and I was obliged to give a speech, as is the custom with visitors, in praise of Atiu. Hardly had I finished this, when a couple of lads with ukeleles and guitars broke into a lilting love song about their island. A love song about a place of misshapen coral, dank caves, and fermented orange juice! Somehow it did not seem inappropriate to my ears.

IF ATIU RESTS AT ONE POLE, AITUTAKI RESTS AT QUITE ANOTHER. OF ALL THE southern, more accessible Cooks, it comes closest to the Hollywood image of a South Seas island. It has all the right amenities: a reef enclosing a turquoise lagoon; uninhabited coral motus; and any number of those proverbially empty beaches. It even gets swept over by the occasional hurricane, a sine qua non for a South Seas island in its Hollywood incarnation.

So it was that I flew to Aitutaki a bit fearful that I might be flying to Tinseltown. But the island allayed my fears from the moment the Twin Otter landed—no, even before that moment, when it *couldn't* land, due to the presence of crab hunters scampering around the runway. Also, it rained off and on, mostly on, during my entire stay, and there's no sweeter, more genuine sound in all the tropics than the patter of rain on corrugated iron roofs.

I arrived late on a Saturday. Next day I decided to explore the island, but that next day was Sunday, and even Aitutaki's ubiquitous mynah birds seem inclined toward indolence on a Sunday. Thus I ended up exploring the island on my own, on foot, in a gauze of warm rain.

First I wandered down to the pocket-size main town, Arutanga. Chief attractions: a harbor that allegedly was Captain Bligh's last landfall before the *Bounty* mutiny, and a cheerful, slightly tumbledown lethargy.

Next I walked inland to Marae Te Mango, a basaltic stone circle honoring, of all creatures, the shark, for it always was serious sharks, not frolicsome dolphins, who guided the old Polynesians from island to island or through difficult reef passages. After paying my respects, I turned north and skirted one banana plantation after another.

At last I came to Maungapu, highest point on Aitutaki. At 407 feet high, Maungapu is no Everest, but it offered me a view that, I submit, could not have been improved upon by Everest itself. I was standing on the summit when all at once the cloud cover lifted, and I saw all of Aitutaki's motus, a necklace of brilliant emeralds strung along a lacework reef. In the lagoon I also saw a solitary outrigger occupied, my binoculars told me, by a seven- or eight-year-old girl masterfully flinging a fishnet into the water.

Later I visited Tunui Tereu, an Aitutaki storyteller, who told me this tale about Maungapu: Once upon a time Aitutaki was only a dull, low-lying place. Locals figured they needed a mountain, so they swam the 161 miles to Rarotonga and hacked off the top of Raemura. (The entire mountain was too big for them to hack off.) This act of vandalism the Rarotongans applauded, since now they had more sun and consequently more time to fish. Holding up the mountain with one of their hands, the Aitutakians swam back to their own island. They installed the top of Raemura just opposite Paradise Cove, where it has remained intact to this very day.

Tunui's wife seemed dubious. "I don't think even Aitutakians could steal a mountain," she said.

I was a bit dubious myself. But I changed my mind after I went to the Rapae Hotel's "Island Night." This included a floor show of lusty, hip-waggling traditional dances, but it also included a local man performing one of these same dances while husking a coconut . . . with his bare teeth. Not since I'd seen the minister on the Scottish isle of Eriskay play the bagpipes and water-ski at the same time had I witnessed a more remarkable display of human talents. Stealing a mountain would be, by comparison, a piece of cake.

The Cook Islands aren't what they used to be, say the *papa'as* (Westerners) who knew them in an earlier era. "Changed, changed utterly, old boy," a venerable Englishman said to me as we sat in Rarotonga's Banana Court Bar, a vintage Polynesian watering hole. But he seemed to accept these changes gracefully, except for one, to him, catastrophic change—namely, the demise of the old copra tubs.

Thirty years ago you traveled from island to island in one of these delightfully squalid vessels or you didn't travel at all. And if you were fortunate, your captain would turn out to be none other than the legendary Andy Thompson, a rowdy, cantankerous, but wholly endearing character who once remarked that he could navigate anywhere in the Cooks simply by looking at the empty beer bottles he'd tossed to the ocean floor.

Maybe the Cooks *aren't* what they used to be in the old days. Or maybe, like Oscar Wilde's Oxford, they never were what they used to be. To chug toward a timeless Pacific isle in a rickety copra boat, with Captain Andy, perhaps drunk, perhaps not, at the helm—yes, indeed, that would have been quite an experience.

But I have a confession to make: For my taste, there are few approaches anywhere on this planet more exhilarating than when your plane sweeps down out of the clouds and, suddenly, miraculously, you see the lush green mountains of Rarotonga rising out of the sea like a landscape from a childhood fairy tale, like a lost world. [1990]

The Marquesas: Beauty, Terror, and Sublime Seclusion

TIM CAHILL

"BUT I RECONFIRMED WITH you, personally," I said. "Three times." The woman from Air Polynesia shook her head sadly. "We have no record of your confirmation."

"But I was here. I talked to you. Two days ago you said I was on the flight."

The woman showed me a handwritten list. There were five names on it and mine was not among them. "You see," the woman said patiently, "we have no record of your reconfirmation."

Outside the office, the island of Hiva Oa sighed under the weight of a heavy tropical sun. In the shade of the mango trees, the air itself was golden, as sweet and thick as honey. The sea was the cobalt blue of deep water, and it glittered in the hard brittle light of unfiltered sun. Breakers eight feet high thundered onto an expansive beach of black volcanic sand. Above the town of Atuona, above the tiny airline office, volcanic peaks rose in immense green velvet spires that caught and collected drifting South Pacific clouds so that rain fell on various summits, and the high peaks shimmered in impressionistic shades of blue and gray.

There are worse places to be stranded, to be sure, but few that are farther from home. Hiva Oa is the southern administrative center for the Marquesas island group, which is about a thousand miles from anywhere: 3,500 miles west of Peru; 2,500 miles southeast of Hawaii; 740 miles northeast of Tahiti. The Marquesas, a part of French Polynesia, are, in fact, the most remote islands on the face of the earth: the island group that's farthest from any continent. This isolation makes tourism rare, and as a consequence, a visit to the islands is rewarding and frustrating in equal measures.

"You could reconfirm for next week's flight," the woman said brightly. There was, of course, no reason to believe that another series of reconfirmations would be recorded.

Ordinarily, in such a situation, I might be tempted to batter my way onto the flight using journalistic clout. However, claiming that I was a big-deal travel writer from the United States would cut no ice whatsoever in Atuona, where there are exactly three tourist bungalows. The accommodations are more a matter of Polynesian courtesy than an attempt to cash in on big tourist bucks.

"You could," the woman suggested, "go out to the airport. Maybe one of the confirmed passengers won't show up."

In another country, I could possibly bribe one of the five passengers: offer a free ticket next week to anyone willing to give up a seat. After spending several weeks in the Marquesas, however, I knew better than to try. The islands are so provident, so rich in game and fruit, that reasonably able folks can live quite well off the land itself. This produces a cavalier attitude toward cash; money is best spent on toys, luxury items, travel. A Marquesan might give up a seat on the plane because he or she liked me, or felt sorry for me, but never for money.

At the airport, all five passengers showed up for the flight. I was destined, it appeared, to enjoy the Marquesas for another week. Or more. There was just enough money left to rent one of the bungalows in Atuona. Oh well, I could hang out on the black beach. Drink cheap rum. Do a lot of writing: hard-hitting, sinewy romantic stuff, full of adventure. . . .

The Marquesas have always been a good place for this sort of creative work. Herman Melville and Robert Louis Stevenson both lived for a time on the northern island of Nuku Hiva. Belgian-born singer and songwriter Jacques Brel moved to Hiva Oa to live the last years of his life when he discovered he had cancer. Paul Gauguin came to Atuona in 1900, looking for inspiration in uncorrupted savagery. Gauguin and Brel are buried in a small cemetery on a hillside overlooking the town of Atuona. Most of the graves there are draped with seashell necklaces and are neatly tended. Brel's stone features a bas-relief of the man. Gauguin's simple grave is headed by a sculpted female figure—Oviri, a disturbing, hollow-eyed symbol of the savagery the painter hoped to find in the Marquesas. A plumeria tree grows beside the figure of Oviri.

At the airport, I found myself thinking about these graves as the small, single-prop plane took off without me. The day I visited the cemetery,

Gauguin's plumeria tree had been in full and fragrant bloom. It had bothered me then that roots surely must protrude into the grave itself. Now, standing alone on a vacant airstrip 740 miles from anywhere, I began to think of those profuse and fragile white plumerias as an expression of the artist's soul. It was a romantic and melancholy thought. Perhaps I would be buried in that pleasant cemetery along with Gauguin and Brel. My tombstone would read: "Tim Cahill: We Had No Record of His Reconfirmation."

THE MARQUESAS ISLANDS ARE VOLCANIC PEAKS THAT RISE TWO TO FOUR THOUSAND feet above the surface of the ocean. Six of the islands are inhabited, and the population stands at about 6,000. Hiva Oa is the administrative center of a southeast cluster of islands, while Nuku Hiva, the main island, is the economic and administrative center of a northwest group. The single weekly flight from Tahiti lands at Nuku Hiva, where the first-time visitor is immediately introduced to the difficulty of travel in the Marquesas.

Nuku Hiva's airstrip is on the northeast shoulder of the island, in a dry rolling grassland that looks something like the wind-whipped high plains of Wyoming, and is called Terre Déserte, Deserted Land, because little that is edible grows there. The main town, Taiohae, lies directly over the spine of mountains that bisects the island, and a dirt road runs from the airport to the village. It is a narrow, dangerous, sometimes boggy affair; a twisting, rock-strewn washboard of a road in which every other turn features heart-stopping dropoffs. The rutted path passes near the highest point on Nuku Hiva, 3,865-foot-high Mount Ketu. It is probably less than 30 miles to Taiohae from the Deserted Land, and the drive takes all day.

Anyone in a hurry takes the ferry, a clattering 60-foot, twin-diesel vessel that docks at a small cove called Haapoli. Boarding the ferry is an adventure in itself. Because the Marquesas are volcanic in origin—the islands simply erupted out of empty ocean—they are not protected by a fringing reef, and the full force of the Pacific Ocean batters them ceaselessly. There is no breakwater at Haapoli, which means that the ferry, docked at a small cement pier, rises and falls as much as eight feet on swells and breakers. Passengers try to step aboard as the deck rises to the level of the dock. To stumble at this point could be catastrophic: A passenger in the water would be crushed between the lurching boat and the cement dock.

It takes about two hours to get to Taiohae, and the boat sails by rocky cliffs, crusted with ancient lava, that rise 800 feet above the surface of the sea. Oceanic swells explode into polished rock. The thunder and spray of high sea meeting solid rock is a constant spectacle: Water rises against the rock in immense sheets, reaching 25 feet or more like moving molten silver under the high tropical sun.

There was no one living above those cliffs on the Deserted Land, and Nuku Hiva was as it always had been, as it was 2,000 years ago when the first of the people who were to become Marquesans landed on the island. They had come out of the west, in dugouts filled with coconut, breadfruit, chickens, pigs, dogs, and heroes. The people, the first Polynesians, ranged as far south as New Zealand, as far north as Hawaii. The Marquesas are the farthest east they came, and the Marquesans' mystic culture—terrifying, brutal, and beautiful—developed in the splendid isolation of these most lonely of Polynesian islands.

As the boat rounds a point by the bay of Hakaui and moves to the windward side of Nuku Hiva, the scenery begins a sledgehammer assault on the senses. The mountains rise sheer, impossibly green, and waterfalls thunder down the drainages, carving out narrow valleys where the people live. The valleys are separated one from the other by high, vegetation-choked ridges. In the time before the first Euro-Americans visited the Marquesas, the people who lived along the streams in those valleys called themselves "The Men." Their coconut and breadfruit thrived in the fertile volcanic soil; the chickens and pigs multiplied beyond counting. Feasting was a way of life—there were feasts for marriage and death and birth—and The Men, to Western eyes, were among the most beautiful people on the face of the earth.

Art was integral to the life of The Men. They carved human images of wood, and filigreed tiny representative scenes on ear ornaments and tortoiseshell fan handles. Males sometimes decorated their entire bodies with tattoos in contrasting bands of light and dark, in swirling circular patterns that followed the contour of a buttock, the curve of the biceps or thigh. There were ritual tattoos that marked the large events of life, such as puberty and marriage. Since the process of tattooing was extremely painful—the colors were made of ground candlenut, ash, and water and were generally applied with a sharpened piece of bone—some body decorations were simply signs of courage: those that swept across the eyelids, for instance, or the genitals.

The most fearsome tattoos served to frighten enemies. The Men were always at war, valley against valley. Groups of warriors would trudge up over the ridges and descend upon those in the neighboring drainage, looking to capture a suitable sacrifice, a man to be killed in the name of the gods. The great stone terraces that The Men built in dark groves above the beach rose to the sculpted stone figures called tikis: large, squatting statues with huge empty eyes and scowling mouths. To consecrate the holy places, the bones of victims were hung nearby, in the branches of the sacred banyan tree. Sometimes, The Men practiced ritual cannibalism.

The first European to set foot on the Marquesas was Spanish explorer Alvaro de Mendaña, who landed at Hiva Oa in 1595 and named the islands for the Marquesa de Mendoza. In 1813, American naval officer David Porter claimed the islands for the United States. President Madison declined the offer, and France declared the Marquesas a part of its empire in 1842.

That was the same year that Herman Melville sailed into the bay of Taiohae aboard an American whaler. In *Typee,* the novel he wrote about Nuku Hiva, Melville described the horseshoe-shaped bay: "From the verge of the water the land rises uniformly on all sides, with green and sloping acclivities, until from gently rolling hillsides and moderate elevations it insensibly swells into lofty and majestic heights, whose blue outlines, ranged all around, close in the view. The beautiful aspect of the shore is heightened by deep and romantic glens. . . . "

As the ferry from the airport enters this bay, the view is precisely as Melville described it, which is to say, awe-inspiring. Taiohae is, arguably, the most spectacular bay in the world. "Very often when lost in admiration at its beauty," Melville wrote, "I have experienced a pang of regret that a scene so enchanting should be hidden from the world in these remote seas." The staggering beauty of the bay bursts upon the eye like privilege.

Taiohae hasn't changed much since the days of Melville. There are no high-rise hotels, no condos or "planned resorts." There are three good reasons for this happy state of affairs. First is the remote nature of the islands. Secondly, land in the Marquesas is generally owned by family groups that may number in the hundreds. To acquire a legal deed, a developer must obtain the signature of every owner. This sometimes means tracking down a cousin who is living in Bakersfield or Borneo or Bimini. Often, by the time the last owner signs, one of the first signatories changes his or her mind. The third good reason that there are only two small hotels on all of Nuku Hiva is the existence of a nasty little sand gnat called the nono fly. The nono's bite is painless, but, over a period of several hours, it raises an ugly red welt that itches like fury. Scratch the welt and it is likely to burst, bleed, become infected. The nonos are a plague, and the lack of condos in the paradise of Taiohae is a tribute to the gnats' vicious ubiquity.

In the town of Taiohae, I stayed in one of the Keikahanui Inn's three bungalows. The inn is owned by an American couple, Frank and Rose Corser, who taught me how to deal with the nonos. A mixture of five parts water to one part liquid bleach is splashed on the body in the shower, then rinsed away. The bleach prevents possible infection and also kills the itch for up to five hours. While in the Marquesas, I generally bleached myself three times a day.

My first Sunday in Taiohae, I went to mass at the Catholic church. Sitting in the pew, smelling faintly of Clorox, I listened to the bishop speak and recalled the passage in *Typee* in which island maidens swim out to the American whaler and do things with the sailors that Melville felt he "dare not mention." The practice of trading sexual favors for such precious items as fishhooks and iron nails was common throughout Polynesia at the time, and the French, through its Catholic missionaries, strove to put an end to the practice.

The missionaries believed that the sins of licentiousness and ritual cannibalism were endemic to Marquesan culture, which they therefore systematically set out to destroy. In this they were aided by the French military, and more significantly, by the diseases they had brought with them. The population of the islands, which probably stood at 50,000 in the 1700s, fell to under 5,000. There was a brief period of revolt against the French—against the death that had come to the islands aboard the big ships, against the killing of the people and its culture—and in the late 1800s, this pathetic revolution erupted into a period of defiant and orgiastic cannibalism. By 1900, the revolt had been crushed, the culture destroyed. Today, the Marquesas are the most staunchly Catholic islands in all of French Polynesia.

After mass, I talked with Bishop Hervé-Marie Le Cléac'h at his home. He was a tall, aristocratic Frenchman who had been in the Marquesas for 14 years. "The people here," the bishop said, "are as they were in the olden time. Nobody obeys chiefs because no one ever did. If the people work together on a project, they do so because they want to and not because someone told them to. They are very individualistic. The motto here is 'mind your own business.' " The bishop seemed to think that this was a good thing.

He was critical of the kind of Catholicism that decimated the culture of the islands a century ago. "Even now," he said, "the Marquesans are governed by foreigners; they learn a language in school that is not their own. They don't know their ancestors, their legends, their history, their art."

What the missionaries had taken from the people, the bishop wanted to give back. "I want them to be proud of their heritage," he said. "I feel I was sent here to make the people proud to be Marquesans. When a man is proud and happy, when he is content with himself and understands his place in the history of his land, then my job is made easier.

He has conquered his confusion, you see, and is then free to accept or reject God as he pleases."

Later, the bishop and I visited Damien Haturau, a handsome and imposing man who is probably the best-known sculptor in the Marquesas. The church had commissioned Damien to carve a life-sized wooden Madonna in the Marquesan manner. He had worked on the sculpture for three weeks, but now he thought there was a fault in the wood, and he told the bishop that he wanted to start over.

Damien lived above the beach in the Meau Valley, sometimes called Sculptors Valley because so many good wood-carvers had chosen to live there. The artist took me into the open-sided shed where he kept his chisels and mallets. On the table, recently completed, was an exquisite carving of an elaborately tattooed Marquesan warrior. Beside the sculpture was a scholarly work on the archaeology of the Marquesas that had been written by a German around the turn of the century. Damien was using the sketches in the book to be certain that the tattoos were absolutely authentic. He was, it seemed, in the process of reinventing his own culture using the archaeology of another.

The bishop studied Damien's carved warrior for several minutes. "Wonderful," he said. "Brilliant, beautiful."

NUKU HIVA IS AN ISLAND THAT BOMBARDS THE SENSES, NUMBING THE MIND WITH its constant and unrelenting beauty. The mountains behind Taiohae rise through groves of coconut and mangoes. A cruel joke of a road threads through papaya, banana, and passion fruit trees. Semiwild goats challenge cars for the right-of-way. A sudden squall blows in from the sea, bringing 10 minutes of heavy rain. When the sun breaks through the clouds, light and shadow race across the green slopes below.

Descending from the central plateau atop the mountains into the valley called Taipivai, our car passed two boys carrying a rooster with a long string attached to one leg. If the boys came upon a wild cock in the jungle, they'd set their rooster on it. Both birds would become entangled in the string and the boys would have themselves another fighting cock.

There are wild horses in the jungle as well, horses originally brought to the island by the French. Marquesans use a mare in heat to draw wild stallions, and the horses are easily broken by riding them belly-deep through the surf. They are small horses, sturdy climbers perfectly suited to the steep, jungled slopes of the Marquesas.

Everyone, it seems, owns a few stallions. In the village of Taipivai, I watched a family of six ride up into the hills, followed by five riderless horses. In less than two hours they were back, and each of the five horses carried over 100 pounds of ripe bananas on its back.

Just above Taipivai, near a wide pool in the clear river that runs through the village, there is a small trail that winds a mile or so through the jungle to the largest of the ancient ceremonial sites in the valley. The great stone terrace, called a *me'ae*, was perfectly square, 25 by 25 feet. Off to one side was a 100-foot-high banyan tree. Aerial roots had dropped from its branches and they had, in turn, become new trunks so that the tree covered perhaps half an acre. The banyan and the thick groves of coconut trees filtered the sunlight, breaking its power. The sacred site was cool and spectral, shadowed in gloom.

A tiki, one of the ancient gods of the valley, squatted before the platform. It was six feet high, but three times wider than any human being. Blue-green algae grew like leprosy across the idol's immense head and obliterated one of the huge, round, empty eyes. The figure was clearly female, its genitals swollen as if in sensual excitement.

It was a simple matter, in the gloom, to imagine sacrifices committed by torchlight, to hear the screams and see the blood flowing over the holy stone. The people of Taipivai, the Typee, had been the most notorious cannibals of Nuku Hiva, the most feared tribe on the island. There had been skulls hanging from the branches of the sacred banyan tree; there had been human sacrifice and ritual cannibalism perhaps as late as 1900.

Later, I sat on a vista point overlooking the valley of Hatiheu and let the afternoon sun bake away the vague, sickly sad sense of dread that had descended on me at the tiki above Taipivai. Hatiheu was the valley where Robert Louis Stevenson had lived. The gentle beauty of the place suited my mood. The hills, alive with soft, calf-high ferns, spilled down toward the sea in waves of variegated green. The sound of the surf rose faintly from the bay below and mingled with the lazy hum of insects, the sigh of the breeze through the trees, the thunder of the waterfalls, and the constant symphony of bird song that was punctuated at odd intervals by a single strangled prehistoric croak.

Below, a series of jagged black rock spires, too steep to carry vegetation, rose like sentinels just above the bay. They might have been the parapets of some alien civilization, or so I thought, and then the idea of blood flowing under the lurid light of torches hit me again so that the contrast of beauty and terror, of life and death, seemed especially vivid.

The tiki had been female, but perhaps the swollen organs were not meant to indicate sexual excitement. Perhaps the god that stood over the place of death was in the process of giving birth.

ON THE ISLAND OF HIVA OA, I STAYED FOR A TIME IN PUAMAU, A VILLAGE OF some 200 souls. As is the case almost everywhere in the Marquesas, there is no commercial hotel in the village. A visitor simply asks to see the mayor, who makes it his business to provide shelter. The mayor of Puamau decided that I could stay in his village, and I was given a comfortable room in his own house, a poured concrete dwelling with a working toilet and shower. Meals were provided along with the room for a cost of about $30 a day. I ate well on roast pork, passion fruit, platters of raw tuna fillets marinated in coconut milk and lime, and French bread that was baked daily in a wood-fired oven.

I had come to Puamau to watch the single most important event in the economic life of the village. The copra schooner was about to make its monthly visit. Copra is dried coconut meat. The islanders collect coconuts, husk them, and leave the meat to dry in large wooden beds. The dried meat is stuffed into burlap sacks and taken to the beach for loading.

The copra is transported by ship to a factory in Tahiti, where the meat is cleaned and crushed. The coconut oil is used in soap, shampoo, synthetic rubber, glycerin, hydraulic brake fluid, margarine, and vegetable shortening.

Copra is the single source of income for most Marquesans, and the arrival of the copra schooner is cause for celebration. People who live up in the hills bring down strings of horses loaded with copra. Everyone gathers on the beach as the boat steams into Puamau Bay. Women play a kind of bingo that may only otherwise be played on Sundays. Young boys strum guitars and sing. Children play in the surf. The able-bodied men hoist 150-pound sacks of copra on their shoulders and trudge out through the pounding surf to the schooner's "whale boat," a 20-foot-long craft designed to carry heavy loads through the breakers and out to the anchored ship.

The schooner is Puamau's link to the outside world. When the whale boat returns from the schooner, it is full of consumer goods ordered by the villagers and paid for with the 30 cents per pound they earn for their copra. I saw the men offload a few stoves, a refrigerator, and some furniture; but for the most part, the copra schooner delivered stereo systems, dirt bikes, motorcycles, and televisions. In the two days it took to load and unload the schooner, I saw at least half a dozen Sonys and as many VCRs invade Puamau.

The French Polynesian government installed television transmitters in Taiohae and Atuona in 1979. Relay stations on the mountaintops beam the signal down to smaller villages like Puamau. The programs are recorded in Tahiti, and broadcast, commercial free, for two hours a day. In Puamau, at the mayor's home, I saw *Towering Inferno*, an episode of "Shogun" and some kung fu epic out of Hong Kong.

A French official, who would rather not be quoted, told me that the government provides television because the people love it. They love it so much, according to this man, that many of them might move to Tahiti, simply to watch "Dallas" once a week. In Tahiti, however, jobs are scarce and there is a growing problem of overpopulation. "The idea is," the official said, "that if Marquesans have TV at home, we won't have to provide food and homes for them in Tahiti."

The fact that television exists in so remote a land, in a place littered with the artifacts of the old times and the old beliefs, sometimes staggers the imagination and presents the traveler with a number of truly remarkable cultural crosscurrents. The day after the copra boat left, for instance, the mayor led me on a short walk through the jungle to a tiki that stood above his house. There are more than 25 tikis in the Puamau valley, but this one was eight feet high: the largest tiki in all of French Polynesia.

Above the ceremonial site there was a steep, rocky spire that rose to a needlelike summit. We began climbing this spire, the mayor and I, using rope he'd brought along. After a stiff hour's climb, we reached the summit. The mayor moved down a few feet among the ironwood trees and reached into a small cave about two feet in diameter. He pulled out a human skull, and began telling a long, incredibly involved story about how a man—or more properly, a man's head—came to be buried there.

It seems that some time ago—I later placed the date at about 1900—there was a queen in Puamau who lived in a house on the large stone terrace not far from the mayor's house. In this time, said the mayor, a prolonged drought threatened the staple breadfruit crop, and there was danger of famine. The queen requested a human sacrifice to appease the gods, and three men took it upon themselves to perform this duty. They pulled all the hair out of their heads and hung their machetes around their necks, which was a dead giveaway to everyone else in the village that they were looking for someone expendable. The people of Puamau gathered together on the beach for mutual safety.

The three hairless men were forced to find a victim in another drainage. "It was over there," the mayor said, pointing to a small ridge, "where they saw a man in a coconut tree." The story became very detailed here. The mayor wanted me to know precisely how many shots were fired (three), and how many times the man was hit (twice), and which was the fatal shot (the one that passed through the right kidney).

The dead man's head and long bones were carried to the top of the spire, where they were placed in the cave. The rains came immediately, of course, and Puamau was saved. The French heard about the murder, however, and sent a gunboat to Puamau Bay. The queen was imprisoned on the ship until the village surrendered the killers. The three men were taken out to the ship at gunpoint and were never seen again. The queen was released.

The mayor pointed down to a house in Puamau and said the name of the family that lived there. "That is this man's family," he said, holding up the skull. The mayor stared into the empty eye sockets, in the manner of a man who is mentally adding a set of figures. "This is the great-grandfather," he said at last.

"And the family knows he is up here?" I asked.

"Yes, of course."

"And they don't want to get this skull and bury it properly?"

"No," the mayor said. He seemed to regard the question as both strange and mildly offensive. "Why would they?"

It was late in the afternoon—the mayor's story had taken almost two hours to tell. He carefully placed the skull back in the cave and said we would have to hurry back down to the house. The television would be coming on soon, and he wanted to watch "Dynasty."

I AM LIVING IN ATUONA, IN A BUNGALOW NEAR THE BLACK SAND BEACH WHERE Gauguin painted many of his most famous oils. The bungalow stands on the site of Gauguin's house. In recent years, the place sometimes has been the home of travelers whose reconfirmations were not recorded. Perhaps these hapless visitors reacted as I did; perhaps they spent a day or two trying to figure out how to change their schedule, or contact the people they were supposed to meet at home. Over the space of days, perhaps others also felt the sun burn away their anger, and maybe they settled into the gentle rhythm of life in the Marquesas.

These days I rise at 6:00, just as the five or six roosters who seem to live under my window explode in paroxysms of earshattering bravado. The old man who lives next door is generally up already, sitting on his porch and playing taped Tahitian laments on his immense JVC boom box.

I pull on my shorts and step out onto my own porch. The old man and I nod to one another, but he expects me to ignore him for the rest of the day, just as I expect him to ignore me. Mind your own business, as the bishop said, is the motto here.

About 7:30, I walk five blocks to the bakery and buy a loaf of good French bread fresh from the oven. I keep forgetting to bring money, but the woman behind the counter recognizes me and runs a tab under the name "M. Américain."

The dog that has adopted me follows at my heels. He strolled into the bungalow several days ago and I shouted one of the few phrases I know in Marquesan: "*Keer aw.*" The words, generally addressed to invading dogs, chickens, horses, or pigs, mean "get out." I had the misfortune of running into a dog who thought his name was Keer Aw. He gave me one of those open-mouthed looks of keen canine anticipation, the kind that seems to say: "Oh boy, wanna play, got something for me to eat?"

So Keer Aw lives on my porch. He sleeps at my feet while I type and tips over the garbage whenever he thinks I'm not looking.

About 11:30, when it gets too hot to work, I walk down to the Gauguin beach for a swim. I am trying to improve my French with the aid of a French/English dictionary and a French comic book I found about the American cowboy, Stormy Joe, and his comical sidekick, Sardine. Already I have learned to say, "Drop your gun or I'll kill you like a mad dog."

Keer Aw lies in the sand while I read. After an hour or so, he gets up and slinks off down the beach, looking back over his shoulder in the most guilty fashion imaginable. "It's not what you think," he says. "I'm not really sneaking back to the bungalow to tip over the garbage."

About 2:30, I walk back home and clean up the garbage. Keer Aw, skulking under the picnic table on my porch, eyes me cautiously. I step into the bungalow, bleach myself down, take a cold shower, and lie down for a brief nap. Outside my window, chickens are lunching on the ants that are feeding on spilled garbage. I find the pluck and cluck of contented poultry curiously soothing.

Some days, I borrow a stallion from a lady who lives nearby and ride over the ridge to a bay called Taaoa where there is a beach that seems devoid of nonos. There is something almost unbearably romantic about riding alone, galloping bareback across the sand as the breakers thunder into the shore.

Some evenings I go to a Chinese family's home on a hill above Atuona's hundred lights. The family runs an informal restaurant. I'm particularly fond of the river shrimp dinner.

Strangely, I've stopped my habit of visiting the airline office and reconfirming my flight every day. I think I'm becoming almost Marquesan in my attitude: If I'm on the next plane out, fine. If not, what the hell. My work is going well, the food is good, the land is vibrant, I'm content, and my dog loves me for no very good reason. Money is no problem. There is plenty to eat growing on the hills above town if it should come to that. Worrying about something I can't control, something like another missed flight, would simply spoil an otherwise perfect day. [1985]

Vanuatu: The Original Bali-ha'i

FRANCES FITZGERALD

PICTURE THE BEACH OF THE travel brochures: the half-moon bay fringed with palm trees, backed by a hillside of dark green jungle. The water is aquamarine and so clear you can see three fathoms down; the sand is white and almost as fine as talcum powder. At the entrance to the bay there is a reef and a line of frothy white breakers, and beyond that a green islet, perfectly centered. In such photographs it is normal, of course, for the photographer to leave out the reason and cause for the travel brochure: a hotel with, say, 168 rooms, a bar serving strange rum drinks, and a dining room featuring weekly barbecues and entertainment having something to do with fiery Hula-Hoops. On the beach I am thinking of, however, there is no hotel; there is no building of any sort and no sign of human habitation. Throughout the day I spent there with a friend I saw no one until the late afternoon, when a man in a loincloth with a spear and two boys with bows and arrows crossed the beach and disappeared into the jungle. They were followed by a man with a radio playing "Hey, Mister Tambourine Man."

The beach I am thinking of lies on a rather large South Sea island. On the east coast there are at least a dozen such beaches, similarly deserted; on the west coast there are more of them, but how many is unclear as there are no roads on that side of the island and the only boats that go there are the trading vessels that sell dry goods and pick up the copra. The island is one of 80 in an archipelago: a chain of coral and volcanic islands stretching 500 miles through the South Pacific. The islands are extremely fertile, and the sea abounds with fish. The islanders live in a surplus economy, so their customary rituals consist of dancing, giving gifts, and drinking a slightly narcotic brew out of coconut shells. The archipelago is also a country: an independent country with a democratically elected government and a stable currency. Its capital is a small, whitewashed town on a clear green bay, its streets shaded by flame trees; it has a century-old trading company, an open-air market where women in brightly colored dresses sell fruit and vegetables, and a restaurant called the Café de Paris. It also has 71 registered banks—only five of which trade locally—and about a thousand corporations in filing cabinets. It is a tax haven. Thus, while foreign newspapers arrive a week late, if at all, telephone and telex services float millions of dollars invisibly—theoretically—through the country each day from Hong Kong, Singapore, and other cities in proximate time zones. Yet few people—at least few people of my acquaintance—have ever heard of this country, Vanuatu.

In some ways this is not surprising. Vanuatu, after all, lies at some remove from the major population centers of the globe, being some 1,200 miles northeast of Brisbane, Australia, and about 600 miles west of Fiji. On the other hand, it is not small for an island nation—it has a landmass the size of Connecticut—and it forms a part of the greater Melanesian archipelago comprising New Guinea, the Solomon Islands, and New Caledonia. During World War II it harbored the largest American field base in the South Pacific, the staging area for the battle of Guadalcanal and the subsequent battles of the Solomon Islands. Between 1942 and 1945 a quarter of a million Americans passed through the base on Espiritu Santo; Bob Hope visited the island; so did Eleanor Roosevelt and the PT boat commander, John F. Kennedy. James Michener celebrated the archipelago in his *Tales of the South Pacific,* and it was the setting for the most famous of all Broadway musicals. From a certain promontory on Santo you can, on a clear morning, see Aoba, the island Michener called Bali-ha'i. Later in the day it disappears in the mists of the islands. The archipelago has, however, disappeared in plain sight, for until 1980 it was called the New Hebrides.

Of course it is in the nature of Bali-ha'is to disappear, for they are, generically speaking, the creation of travelers who see only what they wish to see and then leave quickly before reality intervenes. Because of this solipsism there are Bali-ha'is—South Sea islands of the imagination—floating out across the Pacific from French Polynesia to the Marianas and the Solomons. The reality is rather more various, and for the most part not romantic in the least. It includes military bases and nuclear testing sites; towns made of plywood and corrugated tin; islanders who live exclusively on imported canned goods and welfare checks; young men in Jimi Hendrix T-shirts whose only dream is to get to Auckland or Los Angeles; and young men so angry at the economic and cultural shambles created by one or another metropolitan power that they dream of Kalashnikovs.

Vanuatu, however, is still much as it was when Michener first saw and described it. It is still sparsely populated; there are only 130,000 people on its 80 islands and only two towns: the capital, Port Vila, on the central island of Efate and Luganville on the large island of Espiritu Santo. The population includes a few thousand foreigners and expatriates—Europeans, Australians, Chinese, Vietnamese, and Polynesians—but 95 percent of the population is Melanesian, or, more precisely, ni-Vanuatu. And most people live as they always have, in thatched huts raising pigs, growing taro, yams, and bananas. Vanuatu is not Bali-ha'i, of course. It is a poor country; its main exports are copra, cocoa, beef, and timber. And it is rather more interesting than anything a traveler could dream up.

To begin with, Vanuatu has 110 indigenous languages; given the number of its citizens, that makes the ratio of languages to speakers the highest of any nation in the world. The cultural diversity seems to be equally great—for "custom," or the traditional way of life in Vanuatu, includes an enormous and bewildering variety of rituals, myths, and customary practices. Religions and kinship systems vary not only from island to island, but from village to village. In two or three villages on the island of Pentecost, for example, the yam harvest is assured each year by men who throw themselves off an 85-foot-high wooden tower with lianas tied to their ankles. If the liana is the right length and properly tied, it will break the jumper's fall just before his head hits the ground and will bounce him back up to land on his feet. But not every liana works this way.

On the island of Malakula there is a tribe called the Big Nambas. They are so called because the men wear very large red penis sheaths; the women for their part wear long red fiber wigs. Their neighbors, the Small Nambas, wear small, banana-leaf *nambas* and take pride in making beautiful masks and puppets. On the northern islands men make bracelets out of boars' tusks specially grown so as to describe an almost complete circle of ivory; there, too, there are elaborate grade-taking ceremonies where men attain status according to the number of pigs they can kill and give away. On the southern islands traditional tribal alliances are cemented by a *toka* ceremony in which men and women put on elaborate costumes and dance for days. On Tanna there is a cargo cult that waits for a mysterious figure called John Frum, a man usually described as having bleached hair, a high-pitched voice, and a coat covered with shining buttons; there is also a group that worships the Duke of Edinburgh.

Few tourists ever see these extraordinary goings-on. In the first place, relatively few tourists come to Vanuatu these days; 30,000 in a good year, but it is rarely that many. In 1980 most travel agents lost the country under its new name. The few who rediscovered it were Australian, and it re-emerged on their horizons as a place that could be packaged with New Caledonia. But then the Australian dollar declined and a small war broke out in New Caledonia. According to local hotel managers, many travel agents now believe that there is a war on in Vanuatu.

In the second place, most vacationers who do come to Vanuatu remain on the main island of Efate. And why not? In many ways it resembles a prosperous Caribbean island of 30 or 40 years ago. It has large coconut plantations and cattle ranches and villages clustered around churches and schools. There are several good hotels on beaches and lagoons with windsurfers, sailboats, dive boats, and so on for charter. And there is a country club with a good golf course. Port Vila itself has rental cars, taxis, and nice shops, and its restaurants serve all the local delicacies: lobster, coconut crab, mahimahi, and filet mignon. Vacationers stay on Efate because it has everything they want, but there are other considerations as well. The government encourages tourism on only two of the other islands, and thus while there is excellent regular air service to all the main islands, that is, for the most part, as far as the service goes. The plane lands on a grass airstrip, there is a path leading off into a jungle, and the rest of it is up to you.

Then, too, it is extremely difficult to find any information in Port Vila about the other islands. The guidebooks and travel brochures are not at all specific. As in Greece or India, the brochures proclaim the country a "land of contrasts," but then are quite silent on the kind of contrast you might just run into. "My wife and I went to Epi," an Australian friend told me. "After a bit of a walk from the airstrip we found a beautiful beach. I was up to my knees in the water when I saw an enormous shark heading toward me. Turned out he wasn't interested in me so much as in the weir behind me up an estuary where some people were pulling in fish. Still, swimming was out of the question, so we walked over to the weir to see what was going on. Well, there were a dozen young men in the water—stark naked, of course—biting the heads off fish. Nice chaps, though. They waved and smiled—the blood running out of their lips."

Under the circumstances, one of the best ways to visit the islands is with Keith Barlow, a former bush pilot from Australia who runs his own charter airline. A distinguished-looking man with a guardsman's bearing,

Barlow loves the islands and knows as much about them as any expatriate, having flown among them for 17 years. Flying my friend and me down to Tanna, he put his plane down in low over the large and almost uninhabited island of Erromango to show us the stands of kauri trees in the jungle and the patches of cleared land where islanders had been practicing slash-and-burn agriculture. Tanna is one of the two islands said to be developed for tourism. What that means in practice is that it has a road, a car or two for hire, and an attractive guest house consisting of thatched bungalows, run by a young Australian couple.

Thanks to Barlow, Joe Iaruel, a tour guide and government official, picked us up at the airport in a minibus to take us to the main tourist attraction: a live volcano (one of the five in the archipelago) on the other side of the island. The drive was a long one, and we saw few people along the way except at sundown, when a number of men appeared along the road carrying what looked like the roots of small trees. This, our guide told us, was the kava root, and the men were on their way to a meeting ground under the local banyan tree, where boys would press out the juice of the kava into coconut shells for them. Kava, by all reports from expatriates, acts somewhat like novocaine. It makes you numb around the mouth, then so numb everywhere else that there is little point in trying to walk. The men on Tanna drink it silently, reverently, and then stay where they are for the night.

As it happened, the road had been washed out on one quite precipitous hillside. Iaruel, however, had had the foresight to pick up three young men to sit in the back of the minibus, so that the back wheels held to the rutted track without bogging down or slipping over the cliff. After driving up another hill past a shallow lake with palmettos growing around it, we parked in the midst of a lava field and climbed a few hundred yards to find ourselves on the lip of the volcano, looking straight down into the cone in which were geysers of boiling water and a fireworks display of glowing boulders shooting hundreds of feet up into the air. Constant gnashing and grinding noises came from the center of the volcano; from time to time these were interrupted by explosions—some of them small and far away, some of them thunderous and close enough to shake the narrow lip of lava on which we were standing. Our guide had told us it was perfectly safe, but he had not, we noticed, left the car.

In fact there are all kinds of uncertainties about Vanuatu. The great anthropologists passed the islands by on their way to New Guinea; so, too, did the great travel writers of the 19th century. Thus not very much is known about the people of the islands or their history. From what is known, however, it would seem that both the development and the survival of Vanuatu's extraordinary cultural diversity had a lot to do with loss, disappearance, and solipsism. For this is not the first time the archipelago has been lost. It has been lost repeatedly throughout its history.

It is thought that the islands were first settled around the year 1000 B.C. by people who came from Southeast Asia via Indonesia and New Guinea. The Austronesians, as these people are known, must have been extraordinary seamen, for they colonized not just Melanesia and Polynesia but Madagascar, the Philippines, and Taiwan. At a certain point, however, their migrations ceased, and those who settled Vanuatu forgot or abandoned their naval skills and were lost to their Austronesian relatives for the next 3,000 years. The ni-Vanuatu then proceeded to lose each other—or so the ethnographic evidence would indicate—each tribe or village retreating into solitude to create its own universe. The Europeans first discovered the archipelago in 1606, but then they kept losing it.

The first discoverer, the Portuguese navigator Pedro Fernandes de Quiros, had set out on an expedition financed by the King of Spain to find the great Southern Continent that explorers of the period simply assumed must exist. De Quiros was something of a fabulist—the Don Quixote of his day, it is said. Passing through the Solomon Islands and then sailing south past the islands of Torres and Banks, the ships came upon a large landmass; de Quiros without further ado called it Australia or La Terra Australia del Espiritu Santo. The expedition members went ashore and stayed a month without ever discovering they were on an island. But then the sailors had their troubles: They got along badly with the natives, many contracted severe fish poisoning, and there were several mutinies. Finally the expedition split up, and de Quiros returned alone to the court of Spain to tell fabulous stories about his new continent. But he could interest no one in his discovery, and 160 years elapsed before the archipelago was discovered again.

In 1768 the great French explorer, Bougainville, visited Espiritu Santo, ascertained that it was an island, and discovered several other islands in the group; in 1774 Captain Cook, on his second voyage to the South Seas, made the first marine chart of the islands and called them the New Hebrides—thus placing them in the minds of most people in the neighborhood of Scotland. In the next five years La Pérouse and Captain Bligh followed him. But then the islands were lost or forgotten again for another half century.

In 1826 an English adventurer discovered forests of sandalwood on several of the islands, and after that trading ships began to come in some numbers, for the delicately scented wood was highly prized by the Chinese and could be traded for the tea the British had come to require.

Later the traders discovered another precious commodity in the China trade: sea slugs. In the 1840s a new kind of traffic developed. As the cotton and sugarcane plantations in Queensland required laborers, so traders, or "blackbirders," as they were called, came to the islands to take away men. Between 1840 and 1880 they took away some 20,000 men from the islands, and at the same time introduced firearms, alcohol, and a variety of lethal European diseases. A disastrous population decline ensued, from which the ni-Vanuatu are only just now recovering. Few Europeans, however, settled the islands, for the islanders put up a good deal of resistance to the blackbirding and developed a reputation for ferocity and cannibalism. When the missionaries came they were often mistaken for blackbirders, and were dealt with accordingly. But the missionaries persisted, and until the 20th century they were the only foreigners to take any real interest in the ni-Vanuatu.

There are still a good many missionaries on the islands today: Catholics, Anglicans, Presbyterians, Seventh-Day Adventists, and others. Multiplicity seems to be in the nature of the country, and although most educated ni-Vanuatu are committed Christians—and many of the missionaries are ni-Vanuatu—there remains a certain amount of work to be done. On Tanna, not far from the base of the volcano, we came across a clearing with a few thatched huts and a washline from which hung a set of football jerseys numbered one to 17. This, it turned out, was a mission built by Pastor Joelson Ling and his wife Kathleen, Apostolic missionaries from another island. Kathleen, who spoke excellent English, told me that she and her husband had managed to make 60 converts in the year they had been there, but their task was fraught with difficulty. "Custom is still very strong around here," she said. "For example, people are still in the habit of taking their sick children to the custom doctor to have spells cast over them. Of course it doesn't work. If they would only bring them to my husband, he would cure them. He says prayers over the children, and they get well. He's very good at that."

There were other problems, she continued. "First there's all this kava drinking, and the Presbyterians don't do a thing about it. And then a great many people on this island believe in John Frum."

I had only read about John Frum before coming to Vanuatu, for in the small body of anthropological literature on the islands there were two essays on the cargo cult. During World War II the rumor had spread around Tanna that a strange little man in European garb had appeared to a number of people and had promised to reappear, bringing them boats full of canned goods and modern appliances—generators, refrigerators, and so on—such as the American troops were bringing in in such quantity. Some of the islanders built jetties to receive the ships and hooked up a series of tin cans with string to serve as telephones to call John Frum. While most outsiders believed the islanders were simply waiting for goods (and that Frum was probably the American recruiting officer on Tanna), the cult, according to the anthropologists, had more profound purposes.

Before World War II, the Christian—and particularly the Presbyterian—missions had behaved like medieval churches, running the island politically and economically as well as spiritually. They had pushed people around a good deal and, according to one anthropologist, had broken down the customary political system of the island—a system that consisted of pig-killing ceremonies and wife exchanges along a complicated network of "roads." When World War II broke out, people began to slip away from the Christian villages on the coast and to set up new settlements in the interior, where they drank kava and systematically broke all the other Christian taboos. They also spent all they had at once in the European stores, explaining that John Frum would bring them all the goods they needed, and the European traders would have to go away, there being no more money on the island. The European authorities then arrested the leaders of the movement, understanding it, quite correctly, as a rebellion against European rule.

Since World War II the Christian missionaries have returned to Tanna in great numbers and with more persuasive approaches (e.g., it's OK to drink kava). The missionaries I had talked with elsewhere had assured me that the John Frum cult was on the wane, and that Tanna was now 90 percent Christian. But this, apparently, was not the case.

To go to Luganville on the island of Espiritu Santo is to see that a cargo cult might be the only sensible response to what the Americans did there during World War II. As there was no war in these particular islands, the Americans from a local perspective simply built an enormous base and then left it behind. All around the town of Luganville—indeed for miles into the jungle—there are still monumental remains of the war effort: concrete bunkers, fortified ammo dumps, airfields, concrete platforms, and Quonset huts—some of which are still in use as warehouses. There are even more impressive monuments under the sea. When the troops pulled out in 1945, they gave away all the bulldozers, trucks, and other vehicles the island economy could absorb and drove the rest of them off a point into the harbor. While some of the vehicles were later salvaged, that point—Million Dollar Point, as it is now called—still has an acre of metal around it in the sea.

Snorkeling off that point, I found a very rare crocodile fish that had made its home in a jeep. Not far from Million Dollar Point is the wreck of the USAT *Calvin Coolidge*—the ocean liner turned transport ship that went down on a mine in 1942 with 4,000 troops aboard. As the ship was only a few feet from the shore when it sank, only two men went down with it: the captain and the last sailor on board, whom the captain was trying to rescue. Three football fields long, with its bow at 70 feet and its stern at 240 feet, the *Coolidge* is for divers the largest accessible shipwreck in the world. Going down to look at it with an associate of Alan Power, the Australian dive master who has explored and preserved it, I swam through companionways and watched snappers, batfish, and barracudas glide over a deck covered with aircraft parts and rifle shells.

The town of Luganville—amidst this massive wreckage—has a deserted and forlorn air to it. Only a few cars go up and down its enormously wide, American-built main street. On it there is a Vietnamese hotel, several Chinese dry-goods stores, a couple of banks, and a restaurant with oilcloth-covered tables, where the few expats and foreigners hang out. But the only industry of any size in the town is a Japanese fishing company, its wharf and dry dock not a quarter of a mile from the spot where Michener looked out over the island of Aoba. The Americans have left, but so, too, have most of the French who owned the plantations and ran the economy. The planters left because in 1980 they made the mistake of supporting the forces opposed to independence and creating a small, quite unnecessary, civil war.

The history of the colonial period in Vanuatu reads, in fact, like an Evelyn Waugh novel. In 1906, in the period of *entente cordiale*, the British and the French governments signed an agreement establishing a condominium government over the New Hebrides. The two governments were not concerned with the islands as such; rather they wished to prevent each other's navies from establishing an exclusive claim to the islands, and they wanted to settle the disputes that had cropped up between their respective—Protestant and Catholic—missionaries. The Anglo-French Condominium was a unique experiment at the time, and one which, for good reason, was never repeated.

According to their agreement, the British and the French each had a resident commissioner and a separate system of law courts; thus the French settlers were judged by their own law, the British by theirs, and the Melanesians by whatever law they had the misfortune to run into. Above these separate courts was a Joint Court whose purpose was to resolve disputes between the two groups of settlers and the islanders under their jurisdiction. The first president of the Joint Court, an appointee of the King of Spain, set the tone for the whole enterprise of cooperation. He was, according to one historian, "truly neutral, having understood little French, less English and no Melanesian." But this, the historian remarks, "was no particular handicap, as he was also deaf."

For 50 years the main function of the resident commissioners was to look after their own nationals; only in the 1950s did the Condominium install the services normally provided by governments: health, education, and so on. But then the real trouble began, for while the commissioners agreed to a joint administration of public works, customs, and the post office (jealously watching to see that these departments were equally staffed with their own nationals), they could agree to nothing else, so they created two education systems, two health services, two militias, and so on. The Anglo-French Condominium became quite generally known as Pandemonium. Among other things, it added three new languages to the existing 110: English, French, and a kind of Esperanto-pidgin called Bislama.

In the late 1960s the British and the French, true to form, could not agree on the terms of independence for the islands. The British wanted to clear out entirely; the French, in obedience to Gaullist imperatives in the Pacific, wished to maintain a presence of some sort. Because it was inconvenient to say so, they agreed to the British proposal for elections; then they sent hundreds of French school teachers to the islands in order to create a Francophone majority and set up an anti-independence party with a platform of protecting the Francophones. Their party, however, consistently lost elections to the nationalist (and largely Anglophone) party led by Walter Lini, a former Anglican priest. When independence loomed in 1980, French planters and the French resident commissioner took a fallback position and gave covert aid to a secessionist movement on Espiritu Santo.

The NaGriamel movement was half political party, half cult focused on a tall, charismatic figure called Jimmy Stevens. It had begun as a land reform movement, but in the 1970s Stevens had received financing from a right-wing American real estate dealer from Carson City, Nevada, whose ambition was to create an independent country for himself with no government, no taxes, and an unfettered free-enterprise system. The French had always resisted the movement, but now they supported it. In May 1980 Stevens's men attacked the British compound in Luganville, took over the government offices and the airport, and declared Santo an independent country. The Lini government had no troops to put down the rebellion, and while the British brought in 200 Royal Marines, they could not use them because the French, who also brought in troops, would not agree. On July 30th, in the midst of this impasse, the Condominium and the Lini governments conducted independence ceremonies in Port Vila. The Lini government was now free to act on its own, and it acted decisively: A week later troops from Papua New Guinea arrived via Australian air force planes, took control of Luganville, and ended the rebellion. Stevens was arrested and many of the French planters were politely asked to leave the country.

In Port Vila the Avenue Winston Churchill still runs into the rue Général de Gaulle, but the Lini government has managed to unify the country for the first time in its history. English and French are still national languages, but the official language is Bislama (the word comes from the French *bêche-de-mer*, meaning sea slug), and in the "Republic blong Vanuatu" most people now speak it. There is still a certain unevenness of development (Father Lini has to make his way through UN receptions but also through grade-taking ceremonies, killing pigs), but the government has worked out a modus vivendi between the written legal system and custom law. How its officials do this is not entirely clear in principle, but the settlement of a recent tort case on Tanna illustrates the practice.

The case arose because a custom chief burned down a discotheque—not a discotheque exactly, but a thatched house that some enterprising islander had furnished with taped music and alcohol. The chief burned the disco because a number of young women from his village had gotten pregnant after visiting it—and in his view, *propter hoc*. The local magistrate sympathized with his position, but also sympathized with the proprietor's complaint of property damage. The solution the judge arrived at was to make the fathers of the "disco babies" pay for the damage to the discotheque.

Peace has returned to Santo. Though Luganville has not regained its former prosperity, the remaining expatriates hope that it will. Wondering what happened to Jimmy Stevens's followers, I drove up with a friend to his village of Fanafoa, 20 miles north of Luganville. Passing through cattle ranches and coconut plantations, we suddenly came upon what looked like a custom village from the deep bush. The people in it were naked except for grass skirts, loincloths, and assorted bracelets. The settlement was, however, extremely orderly and clean: Its thatched houses lay in straight rows with swept earth in between and a clipped hibiscus hedge surrounding them. A haunch of beef hung from a tree. As we asked directions in sign language from a man carrying a machete, I noticed that the young women wore skirts not of grass but of croton leaves, beautiful aprons of red, yellow, and green.

Continuing down the road, we came eventually to several concrete buildings. One of them was a school and another the former headquarters of the NaGriamel Party. One of Stevens's daughters and her husband lived not far away in a thatched house surrounded by fruit trees; in their yard was a 1941 Dodge army truck, still in working condition. While Stevens's daughter served us pineapple juice from a big pitcher,

her husband, James Tangis, told us something about the village. The people we had passed, he said, were The Naked People, a tribe that had come out of the mountains in the 1960s and attached themselves to Stevens. (The tribe, I read later, had belonged to another cargo cult. They were naked because they preferred it that way, having considered the alternatives.) Apart from them, he said, most of the people in the village were Christian, as was he. They raised peanuts as a cash crop and ran some cattle. But the village, he said, was not what it once was: There used to be many more people, acres of lawns, and a radio capable of broadcasting to the whole archipelago. He had been the radio operator for the NaGriamel and had spent a year in jail after the movement collapsed. "We are all right here," he said, "but we are just making do. We are waiting for Jimmy Stevens to come back. They say it won't be too many more years now."

Later he took us down behind his house to a river that flowed through a Gauguin jungle. His three children were catching freshwater shrimp with snorkels in a clear, deep pool, and just downstream two other children were sunning themselves on a fallen tree.

We went back to town, and later an Australian couple, Eddie and Pauline Beljaars, took us out on a motorboat to the resort they had bought and refurbished on a tiny island at the entrance to Luganville harbor. Bokissa Island had a white beach with aquamarine water, a lovely coral reef, and a forest with a kauri tree, a banyan, and a tree called a navaru, which blooms at night in a burst of white and pink flowers. The resort itself had luxurious bungalows, an excellent chef, and a telex machine. But the Beljaars had no other guests. They were waiting, they said, and hoping for the Australian dollar to rebound and for the tourists to come back to Vanuatu. [1986]

Great Barrier Relief

CHRISTOPHER BUCKLEY

"YOU MIGHT SEE SOME MANta rays," said the pilot of the 1953-vintage Beaver floatplane. "We've been having them lately, great big, black triangles, some 15 feet across." Six hundred feet down, the water was deep blue and shimmery. I couldn't make out any black triangles. Twenty minutes later the pilot set the Beaver down with a *bump-bububbub-bum* off Hinchinbrook Island, a wild, woolly—and scaly—piece of land within the tropical islands of the Great Barrier Reef, and the first hop on a tour that followed, in a loose sort of way, along the watery footsteps of Captain Cook.

It was no small trail. The Great Barrier Reef stretches for 1,240 miles off the northeastern coast of Australia. It's been called, variously, the Eighth Wonder of the World and the "World's Largest Living Thing." To Cook it was probably the "World's Greatest Bloody Nuisance." He ran hard aground on it, on June 11, 1770, and history would have been poorer had he not succeeded in getting off.

The Portuguese were probably the first Europeans to sight the reef, but Cook was certainly the first to experience it up close, during the first of his three famous voyages. Capt. Matthew Flinders, who came in 1802 to fill gaps in the charts, may be responsible for the term "Great Barrier Reef," but the name is actually something of a misnomer. The reef consists of 2,100 separate reefs, scores of vegetated cays, and 540 high continental islands (that is, islands that were once mainland peaks until a great melting more than 6,000 years ago caused the water level to rise 500 feet).

It can be justified as the World's Largest Living Thing—no offense to the blue whale—not by virtue of the 2,000 species of fish or thousands of mollusks, sponges, worms, crustaceans, and echinoderms that inhabit it, but by its being made up of coral, more than 500 kinds, soft and hard—definitely hard enough to punch through the keel of Cook's bark, HMS *Endeavour*.

Hinchinbrook Island, which Cook spied three days before going aground, is one of the largest island national parks in the world. Mountainous and dense with rain forest, it's been described as a 250-square-mile chunk of New Guinea that broke off and floated south across the Torres Strait. Cook was convinced it was part of the mainland, a rare mistake for that navigator, and ironic considering it was he who determined during the trip that New Zealand was not, after all, the eastern coast of the "Great Southern Continent"—Australia. The first written reference to Hinchinbrook as an island does not occur until 1843, when naturalist J. Beete Jukes, sailing aboard HMS *Fly*, called it a world unto itself and one teeming with life.

Teeming, by one current tally, Hinchinbrook is host to: six types of frog, one of crocodile (the Australian saltwater variety), four turtle, three gecko, seventeen lizard, eleven snake (more on this in a moment), five marsupial, six bat, five rat, one marine mammal (the dugong), and 149 kinds of bird.

Wilma, the pretty, Ulster-born assistant manager of the resort whose motto is "Max. Pop. 30" gave us a briefing over a barbecue lunch of bangers and jewfish: "You'll see a wallaby (a short version of the kangaroo) or two on the island. Our favorite is Mr. B, and we ask you not to feed him anything with refined sugar, as wallabies' systems cannot tolerate it, and they go blind. We also have rather pretty bush rats; you'll see them everywhere, and lizards called sand goannas. We advise keeping your screen doors shut, as they do love the fruit baskets."

An interesting concept in hospitality management, I thought: Let the guests fight it out for the complimentary fruit baskets with giant lizards.

"There are a number of snakes on the island," Wilma continued, "some of them poisonous, such as the Death Adder . . . "

"Breath Adder?" I said through a mouthful of Hinchinbrook surf'n'turf.

"Death Adder. You'll see one over there in a jar by the bar. Most poisonous snake in Australia, but we've never had anyone bitten yet, so I don't suppose you two will be the first we lose." Smile.

Along for the journey was my oldest friend Daniel Gilbert, free-lance explorer and jack of many trades. That night he and I set up a defensive perimeter outside our bungalow, a sort of jungle version of a trailer. We were inspired to do this by the sounds coming from behind the bathroom mirror; it had to be one of the island's plentiful white-tailed rats, we decided. (Wilma calls them "possums," because of their size, I think, and because it sounds better.) Outside, there was a loud crash in the bush. We quickly had our flashlight beams on it. Mr. B. Eventually he hopped over and accepted some apple halves, which he ate slowly, with a loud chomping sound.

Hinchinbrook is one quarter the size of Rhode Island, and of penetrability ranging from good to un-. We had no hope of seeing it properly during our short visit. Not that we weren't grateful for the roof over our heads—which made a pleasant sound during heavy morning rains—but the way really to see Hinchinbrook is by boat, through the labyrinthine, 30-mile-long channel separating it from the mainland. Or by backpack. It would take four days to trek from the north end of the island to the south. There's a waterfall at the head of the mangrove swamp at Zoe Bay, and, near the summit of 2,900-foot Mount Straloch, a B-24 Liberator bomber.

On December 18, 1942, the plane—nicknamed the "Texas Terror"—took off from Townsville in an electrical storm with seven passengers and a crew of five. What happened to her remained a mystery for 11 years, when the wreckage was found. In 1960 a group of local residents placed a memorial cross, fashioned from the wreckage, at the crash site.

We missed the plane on this trip, and the trek across the island. (The park rangers in nearby Cardwell who issue the required camping permits warn against pitching a tent along the banks of a certain creek, unless you want to provide a midnight snack for the estuarine crocodiles.)

But we did follow a short, marked trail through the rain, paperbark, and eucalyptus forests and saw giant strangler figs, guandongs, bloodwood and ironbarks, pandanus and swamp box. We came to an overlook and saw the nearby Brook Islands, a nesting place for turtles and pigeons, and a more hospitable place now than during World War II when goats were shut up in bunkers and bombarded with mustard gas to see how they did. (Not great, I have a feeling.) We stumbled along in the rain forest darkness, eardrums rent from the screech of sulphur-crested cockatoos.

Later that afternoon, toward sunset, I was fishing from the jetty and saw a gray shadow moving slowly through the water, too slowly for a shark. I can't claim to have heard it break water to make its soft, trademark sigh, but I think it must have been a dugong, close relative to the manatee. The dugong is a gentle mammal that the aborigines once hunted here along the shores of Missionary Bay, named for a 19th-century man-o'-god named Fuller who came here determined to convert the heathen. He set up camp on a beach, stayed for six months, and never saw a single person. They were watching him at a safe distance, observing him at his noisy prayers and strange gestures. They decided he was crazy and to be avoided.

Only bushwalkers came to Hinchinbrook before the 1960s. Then in 1966 a small resort was planned as an escape from the tourist meccas starting to crowd the Queensland coast. But the island had a way of resisting developers who announced lofty environmental goals in one breath and in the next set about bulldozing an airstrip, quarry, and garbage dump. They went bankrupt. When the present resort opened in the mid-1970s, a seaplane carrying the first load of guests crashed on touchdown, just beyond where I saw the gray shadow prowling the shallows. Everyone got out alive, but the spirit of the islands had made its point.

IT'S 25 MINUTES BY BEAVER FROM HINCH TO ORPHEUS. WE TAXIED UP TO THE "Orpheus Island International Airport," a 15-by-10-foot float a few hundred yards off the beach. That's about as close as you can get. The rest of the way is by landing craft. As Henny Youngman would say, when the tide goes out at Orpheus, it really goes out—exposing acres of dun-colored coral.

Coral polyps sustain themselves symbiotically with the single-celled algae called zooxanthellae. Every year a single acre of the polyps excretes between 12 and 24 tons of limestone onto the walls and foundations of the reef. At low tide they also secrete their own version of SPF 30, a mucus that protects them from the sun. When it gets on humans, however, the mucus can prevent cuts and scrapes from healing.

Nikki Clark was filling us in about all this when her husband, Alan, the general manager of the resort during our stay, came over and told us about a helicopter crash the day before. A Bell Jet-Ranger carrying tourists on a "joy ride" from Dunk Island to Palm Island developed engine trouble and made a forced landing off Orpheus. "Luckily they all got out," said Alan, "but there were some serious injuries."

"It's tricky getting out of a crashed helicopter," Nikki added. "You have about ten seconds before it sinks, but the blades are still turning, so it's a choice between drowning or taking your chances in a Mixmaster." Pleasant thought.

Orpheus owes its lyrical name not directly to the Greek who kept the Argonauts safe from monsters with his music, but to the memory of HMS *Orpheus*, a steam corvette that went down with all 188 hands off

New Zealand in 1863. By contrast with Hinchinbrook, 3,200-acre Orpheus is relatively small and barren. The only snake we saw on the island was a child's harmless python curled up on top of a Haig whiskey bottle over the bar.

There were the usual million or so cane toads hopping about everywhere, but also one brilliantly green toad inventively named "Frog," who inhabited a corner of the open-air lounge next to the picture of the Duke and Duchess of Something. (Orpheus seems to be *Terra Celebritata*: "Elton loves it here, you know"; "I had a call from Paul (Crocodile Dundee) Hogan's secretary this morning, wanting to know if we could get two giant clam shells for him and his wife to use as wash basins. Wouldn't *that* be lovely?") The mango trees behind my room sounded like a hangout for a significant number of the island's fairly copious fruit bat population.

Finally there are 4,500 goats on the island. Divided by 3,200 acres, that comes out to too many goats. The population is the wayward result of a goodwill gesture that began around the turn of the century, when four of the animals were put on the island for the benefit of shipwrecked sailors.

The James Cook University of North Queensland operates a research station in Pioneer Bay, on the western side of Orpheus. They had 72 different projects going, but the one we wanted to know about concerned the creature that, of all the ones we'd seen so far underwater, fascinated us the most: *Tridacna gigas*, the Pacific giant clam.

The manager of the station is an engaging man named Geoff Charles, whom I'll resist calling Clamadile Dundee. He's a former ship's master, former just about everything associated with the reef. He talks of the days when he was one of the hunters, rather than the protectors, of marine life, in the demonstrative tones of someone who's been through a Vietnamese re-education camp: "We were guaranteeing 1,200 fillets a week, and we were damaging these bommies (coral heads) with our anchors, destroying the coral. But I didn't know it at that time."

There were hundreds of the clams, all resting in large blue-green plastic tubs of Geoff's biologically sterile saltwater—"I'm world famous for my filtration systems, I don't mind saying"—magnificient bivalves with purple mantles coated with life-giving zooxanthellae, and studded with thousands of photosensors encircled by electric green iridocytes. Aging, impotent Formosans once craved the abductor muscles of these clams for their supposedly stimulating effect, and kept afloat a fleet of Taiwanese poaching vessels to produce them.

Where giant clams used to provide some of the staple diet of a number of Asian and Pacific countries, now they're classed as an endangered species. Thousands of juvenile clams are shipped from the Orpheus Island station to replenish depleted stocks. "This batch here," said Geoff, "is going to Fiji."

He spoke about his brood with pride, affection, and wonder, especially about their medical applications in curing allergies and transplanting tissues.

"Look at that one," Geoff said, pointing to a young clam about seven inches across. A small chunk was missing from its lip. "Poor thing got chewed on by a turtle when it was out in the bay. We'll put it right."

ON OUR LAST AFTERNOON AT ORPHEUS WE HIKED THROUGH GROVES OF ACACIA, our ankles stinging from the gummy tips of spear grass. We made our way through fields of windbent eucalypti, over granite boulders splotchy with bright orange lichens. Climbing down a canopied gully, we grasped at leaves for support and came away with hands covered by stinging green ants. We walked along mangrove swamps loud with the popping of pistol shrimp and the crepuscular shrieking of sea eagles. Then, near the top of Fig Tree Hill, we came to a gravestone piled high with a strange collection of offerings—shoes. The inscription read:

In Memory of
Mr. Nicholson
A Black and White
Fox Terrier
Who Sailed These Waters
On the Wandoo
From 1953 to 1966
Fidelis Canis

Mr. Nicholson perished from eating a cane toad. Since he was never without a shoe to chew on, they placed on his grave the one he'd been working on when he died. Not long after, another shoe was put there, then another, and another, and eventually the idea took hold that his spirit would grant good luck to any who left a shoe.

By the time we arrived, almost a quarter-century later, the grave was piled high with shoes—sneakers, sandals, even a pair of patent leather shoes worn by a visitor to the island on his wedding day and mailed back to Orpheus with a note asking that they be placed on Mr. Nicholson's grave.

GREEN ISLAND IS A "TRUE" CORAL CAY WITHIN THE REEF, FORMED BY MILLIONS of years of accumulated sand from dead coral. Cook may have landed here; that history is unclear. Also unclear is the origin of its name, which may derive from the Royal Society astronomer who was aboard HMS *Endeavor* to carry out the ship's overt mission: to sail to Tahiti and observe the transit of Venus in order to fix the distance of the earth from the sun. The *covert* mission was to continue westward "to make discovery of the Continent . . . that there was reason to imagine existed." Cook found it and charted more than 5,000 miles of the east coast of this great continent of "New Holland." Poor Green died on the way home, so it seems right to honor him by thinking of the island as his.

Unfortunately, Green Island is to Cairns—the nearby city on the mainland, with its international airport, large hotels, and package tours—what Martha's Vineyard is to Massachusetts—very accessible and highly trafficked. It's a beautiful island, no doubt about it, but its beaches swarm with day-tripping teenagers drunkenly shouting the theme song to "Gilligan's Island." And there's something truly depressing about watching a noble white heron on top of an outdoor table picking at hamburger wrappers and greasy french fries, looking at you with a pickpocket's eyes. But when the tide is out and the reef is laid bare, you can walk, in sneakers, all the way out to the edge of the water, taking in along the way the abundant marine life in tidal pools and other coral nooks and crannies.

The best reason of all to go to Green is to see Marineland Melanesia, a superb aquatic museum run by a gentle man named George Craig who, before he fetched up on this shore, spent 20 years hunting crocodiles in the swamps of New Guinea.

Here, beyond the bridge over the pool with the black-tip sharks, are "Oscar" and "Cassius," two of the largest crocs in captivity. Every morning at ten, George dangles a pig's head for them from the end of a pole. This is a spectacle not to be missed, especially if there's a group of Japanese tourists on hand for a sound track. Cassius is about 18 feet long—with crocs, measurements tend to be approximate, unless you care to jump in with a tape measure—and an alarming amount of him rises up out of the water for breakfast, leaving you resolved never to venture into croc water in anything smaller than an aircraft carrier.

George told me the story of Cassius on the boat back to Cairns. (He was going to the supermarket to pick up some pig heads.) Cassius was languishing at an outstation of a cattle ranch on the other side of Australia. He was, George said, "a real old tiger," almost a hundred years old, scarred with a bullet hole or two, missing a leg, and harassed by onlookers who threw stones at him to make him stir. George got Cassius trussed up on a pallet and drove him 4,000 miles across Australia to Green Island, where we saw him, lying in the midday sun in a pool of jade green murk and lily pads, emitting through his armored snout a faint, contented hiss.

"FOUND," WROTE COOK IN HIS LOGBOOK THE NIGHT OF JUNE 11, 1770, "THAT WE had got upon the edge of a reef of coral rocks." We could see the spot where it happened from our seats, a thousand feet in the air, aboard the twin-engine Otter that was carrying us from Cairns to Lizard Island.

Indeed, the whole critical trajectory of Cook's transit of these fraught waters was visible from the window. Behind us was Cape Tribulation, so named by Cook "because here begun all our troubles." Below us now was Endeavor Reef, where in 1970, 200 years after the incident, salvagers raised the six cannons the captain had ordered thrown overboard to lighten his stricken vessel. To the west we could see Weary Bay, into which *Endeavor* had come after its anxious 23 hours on the reef, seeking a place to careen for repairs. Up the coast we saw Cooktown, at the mouth of the Endeavor River, where sailors patched the hull. North of Cooktown we could make out Cape Flattery, which had "flattered" Cook into thinking it was the continent's northernmost point; and beyond that, just visible, Lookout Point, from which he spied a group of "high Islands" offshore, which he resolved to "Visit . . . in my Boat, as they lay at least 5 Leagues out to sea and . . . from the top of one of them I hoped to see and find a Passage out to sea clear of the shoals."

On the afternoon of August 12, two months after the grounding, he and Joseph Banks, the expedition's naturalist, landed on the beach and climbed to the top of the 1,200-foot-high granite mountain and saw their passage, ten miles distant, out through the reef to clear water. "The only Land-animals we saw here," Cook wrote in his journal, "were Lizards and these seem'd to be pretty plenty which occasioned my nameing the Island Lizard Island."

We heard more on Lizard about Elle Macpherson than we did about the great navigator. The famous model and not one but two camera crews from *Sports Illustrated* had just been there for the magazine's annual swimsuit issue, along with enough suits to fill an entire suite—500 of them.

The resort on Lizard was begun in the sixties by a Queensland bush pilot named Syd Williams, who used to fly out for the black marlin fishing—probably the best in the world. Now it's owned by Australian Airlines. It calls itself "The Jewel in the Reef." The island itself is breathtaking, with a blue lagoon and 24 beaches, and the resort is a member of Relais et Chateaux, employing seven chefs among a staff of fifty. There are only 32 suites. When the black marlin are running, between August and December, fishermen from all over the world descend on the island. One night, poring through scrapbooks, I came across a faded color snapshot of Lee Marvin, a rare smile on his face, fresh from the hunt. Over the years Halloween has evolved into a mad scene here; last year 300 people showed up for the party.

The evening we arrived they were showing videos of that day's diving trip to the "Cod Hole," one of the hot spots on the outer reef where dozens of huge potato cod congregate to be hand-fed by divers. Lizard has some of the clearest and bluest water and the best diving of the islands, because it's so far offshore and therefore near the outer reef. That day a 20-foot-long minke whale had shown up at the Cod Hole. No one tried to feed it.

The next morning we walked over the granite hump known as Chinaman's Ridge, across the mangrove boardwalk officially opened by Prince Charles, to the ruins of Mrs. Watson's cottage on the long, white strand that bears her name.

Mary Watson was the wife of a fisherman who lived on the island. He went off to fish for sea cucumbers and left her with their three-month-old baby boy, Ferrier, and two Chinese servants, Ah Leong and Ah Sam. When aborigines attacked, Ah Leong was killed, and Ah Sam took seven spear wounds.

On the afternoon of October 2, 1881, Mary placed Ah Sam and the baby in an iron boiler and paddled away from the island. It is painful to read the journal of their nine-day ordeal that was found three months later on Howick Island, 35 miles away.

Oct 7— . . . *Saw a steamer bound north. Hoisted Ferrier's pink and white wrap but did not answer us.*

Oct 10—*Ferrier very bad with inflamation . . . self very weak; really thought I would have died last night.*

Oct 11—*No rain . . . Ah Sam preparing to die . . . Ferrier more cheerful. Self not feeling at all well . . . No water. Nearly dead with thirst.*

Her remains and the baby's were found by the captain of a schooner, huddled in the boiler, which, ironically, was half filled with rainwater.

I put my hand on the ruins of her stone cottage and tried to conjure an image of her and Ferrier and Ah Sam going about the business of living an ordinary, agonyless day.

We made a scuba dive that night at a hole not far from Mrs. Watson's beach. This was my first night dive, and I confess to mixed feelings about it. Sinking to a depth of 65 feet in black water frequented by 15-foot hammerhead sharks is not really my idea of a perfect night out. About 40 feet down, while I was making an anxious, 360-degree protective swivel, I looked through my already fogging mask and saw a five-foot-long loggerhead turtle. He was rowing away from me, and we were probably both grateful to be going in opposite directions. With my flashlight and phosphorescent green Cyalume stick, I must have looked like some weird apparition, an ugly, noisy underwater firefly disturbing his slumbers for no reason other than purposeless curiosity.

The next morning I hiked alone up to Cook's Look. On my way to the hill I saw a one-legged man come ashore in a dinghy on Mrs. Watson's beach and hobble along with a piratical gait.

The trail was steep in parts. I pulled myself up by grasping eucalyptus branches and strangler vines. There was jasmine and pea-flower, and a blue-spotted butterfly flitted past my nose. I tried to crowd out thoughts of airline reservations and seat assignments by wondering if Cook and Banks had slipped on the sharp rocks with their heavy-soled leather boots.

Below, Mrs. Watson's cottage dwindled to a speck. I reached the summit—though that sounds overdramatic for a 1,200-foot peak—an hour and a half after setting out. The wind was blowing stiffly, 30 miles an hour out of the southeast, flattening the kangaroo grass.

I tried to find the passage through the reef that had revealed itself to a desperate Cook that morning in 1770, but I couldn't in the haze. So I added my small stone to the large cairn that has piled up over the years and wrote my name in the visitors' book beneath that of Lauren Wrestley ("*8 years old—Thout it was boring*") and started back down to begin the long trip home. [1989]

New Zealand: Hard Cases and Room for Hope

MICHAEL PARFIT

THE AIR WAS CALM. THE little airplane floated through it, droning along in peace. Ahead, a long white cloud made a roof over an innocuous haze. The haze was pale blue, as clean as the waves below it. The shore was dark under the cloud's shadow. Far up in the haze, the shape of Mana Island began to appear just off the mouth of Porirua Bay. The island was a faint brown silhouette, suspended in the haze and shadow.

Everything seemed so still. I was lulled. I let the hand on the wheel rest, and gazed out the window. Ahead, Mana Island stood on the edge of vision. Mana: the Maori word includes courage, authority, wisdom and perhaps humor—all those things to which a man or a woman or a nation must aspire—in the idea of power. Like the meaning of the word, the island ahead was huge and indistinct. It was a slab of iron on which the wind and the sea hammered without effect; but it was as light as the air itself. As I stared ahead, the shape of Mana grew. The full name of this island is *Te Mana-o-Kupe-ki-Aotea-roa,* which means, roughly, the ability or power of Kupe (the original Maori explorer) to sail to the Land of the Long White Cloud. It was indeed a benevolent power, the mana of Kupe, that brought him to these islands, but just what kind of power that Polynesian word describes is unclear. Everyone in New Zealand uses the word, but no one describes its meaning in quite the same way. Influence, power, authority, stature: mana has an elusive meaning, but it haunts this gentle land.

The plane swept under the eave of cloud, into the soft haze, and the air went wild. The airplane leaped and shuddered. The baggage shifted in the back. The altimeter needle fell 50 feet with a thud. The ceiling speaker crackled with the calm voice of a man who was safely on the ground. "The wind at Wellington is three-four-zero degrees at 35 knots gusting to 50," he said. "For Wellington, that is considered normal."

I eased off on the power, and let the plane buck. But in the tumult, Mana Island, with all its cargo of meaning, disappeared aft, forgotten in the overwhelming urge to land, to get my feet once more placed securely upon the good, kind earth of New Zealand.

New Zealand is a comma of land at the end of the earth, separating the southern Pacific from the Tasman Sea, and when the westerlies blow, they whip across the country without a pause. Unlike the Maoris who arrived in canoes, I had come to this nation of islands from America as most *pakehas*—white men—do, in a swifter vessel, a 747.

As we had landed yesterday in Auckland white clouds drifted overhead in the wind, recollecting the Maori name for New Zealand—*Aotearoa*, or Land of the Long White Cloud. There had been wind in Auckland, too, unfurling dark squalls across the water of Waitemata Harbor. When I rented this little plane to fly from that largest city of New Zealand to Wellington, its capital, 400 miles south, the wind still assaulted the coast.

I was now caught by the end of the lash. The plane bounced through the aerial maelstrom at the south tip of the North Island. Most of New Zealand's 3.2 million people live on North Island, but not many live right here, where Mana Island and Cape Terawhiti catch the fury. The land below was treeless and empty. The cabin speaker crackled. "Wind at Wellington now three-four-zero, 35 gusting to 45."

What an arrival! Wellington, home of 320,000 hardy souls, gets 118 days a year of winds greater than 40 miles per hour. For Wellington, normal. A friend told me that in his garden he had to stake down his cabbages. His great unrealized ambition had always been to grow a pumpkin that did not get blown away before its time.

In the streets of Wellington, the wind blew. Wind surfers used it to whirl across the harbor. Phalanxes of barrel-chested men in shorts jogged solemnly into it, getting fit for rugby, the national sport. Downtown, it made the tall buildings creak and sigh. It hummed in the huge fins of the building New Zealanders call the Beehive, which houses the nation's government offices.

Government is such an overwhelming presence in New Zealand that the Beehive must be symbolic. But the ubiquitous government—which requires that you wear your seat belts, demands that your dog be licensed, pays your doctor, and taxes you at around 60 percent if you net more than $38,000 New Zealand dollars a year (about US $27,000)—is generally so benign and friendly that the strangeness of the Beehive must surely proclaim that New Zealanders don't entirely take their government seriously.

The Beehive is a roundish edifice with fins that stands beside the sedate, elderly Parliament building. No one seems to know exactly what the Beehive cost, but it was probably too much. The Beehive dwarfs the Parliament structure, and is connected to it by a vast, above-ground duct. When you stand outside this assemblage, out by the statue of Richard John Seddon, one of the architects of New Zealand's welfare state, there is no escaping the fantasy that the Beehive is really a huge air conditioner for a rather small building that must be the devil to cool.

I left Wellington in a brief, rare moment of relative calm. "Wind One-Six-Zero degrees at 10 knots." South Island came at me slowly: capes and mountains. Only 852,000 people here; this is the island of space. I left small farm fields and a glittering city on the shore of North Island and found on the other side of Cook Strait a place of empty hills and small villages widely spaced along the beach.

I flew down the east coast at 1,500 feet. The twin ranges of the Kaikoura mountains were smothered by clouds. Through breaks in the overcast I saw tiny patches of snow. The land was brown, still in the grip of the drought that has been borne across New Zealand on the endless westerlies for 18 months. Small thirsty rivers ran swiftly out of the hills, loosening their braids as they came to the sea. The Waima, the Clarence, the Hapuku, the Kahutara. At last, a few miles north of a long beach of black gravel pointed straight at Christchurch, 100 miles south, I could no longer stand being above all the loveliness. I turned back to Kaikoura, and landed.

I rented a cabin near the plane, then hitched a ride for the four miles into Kaikoura. The clean little cottages of the town had roses and hydrangeas in their yards, and signs out front: "Fresh Veges, new potatoes." On the city beach I watched the surf rattling stones until a young man bearing his own symbols of mana, a heavy red jacket and a motorcycle helmet, hastened from the road to the edge of the water, where he stood staring out at the sea as if desperate for the relief of its grandeur. I left him to his abrupt contemplation, found a restaurant and ate a crayfish—the large, lobster-type delicacy of the sea for which the town is named.

In the long summer dusk, I walked back on the beach. The land and the sky were gray, the shades of the new world before color came. It was a moment of perfection. But then, New Zealand has been Utopia for centuries. To Edward Gibbon Wakefield, the force behind the New Zealand Company—which settled chunks of both islands in the middle nineteenth century, New Zealand was to be the canvas for a new "Art of Colonization." His experiment would avoid the pattern of settlement he thought turned other colonists into a people who were "rotten before they are ripe." When Captain William Hobson came in 1839 to establish British law in New Zealand, and to mediate with the Maoris, it was with a new kind of imperialistic mandate, one that sought to avoid the "process of War and Spoliation." When Richard John Seddon was prime minister at the turn of the nineteenth century and his radical cabinet was making industrial arbitration laws, creating old age pensions and giving the vote to women, New Zealand was thought to be the land of the future. No one knows exactly what the adventurers in the legendary Maori canoes of the fourteenth century expected of the place, but it drew them here from their unknown past through what must have been daunting peril. To Sir Peter Buck, the noted Maori scholar, New Zealand could be the place where, for the first time on earth, two races mingle their strengths to make a being who could become "an outstanding figure among the differing races of the world." In New Zealand, there has always been room for hope.

I slept well in the little cabin, with the window open to hear the waves. To me, Kaikoura was Utopia enough. But the morning came, and I had to leave.

Again, New Zealand slid under the wing. That straight beach, villages on the shore: Goose Bay, Oaro, Claverly, Conway Flat, Hurunui Mouth. The mountains retreated to the west, the land opened out, the decaying volcanic ruins of the Banks Peninsula rose like a brown shadow in the haze to the southeast, and I was on final approach to Christchurch. The great Canterbury Plains spread out for 100 miles to the south, endless farmed acres, divided by dirt roads and long, straight irrigation canals full of turquoise water. The water was diverted by little yellow canvas dams out into the fields, where it seemed to turn instantly green.

Christchurch was founded by what Edward Gibbon Wakefield called "not merely a nice, but a choice society of English people." It remains the most British of New Zealand towns. On either side of Christchurch, vast wild rivers, the Rakaia and the Waimakariri, cross the plains in gravel washes often more than a mile wide, but in Christchurch the Avon flows under arched bridges and between green banks, under the arms of willow, chestnut and oak. The Maoris believed that anyone who could change nature—bring on the thunderbolt or alter the course of the wind—was rich in mana. Here the force of yearning, the desire of the choice society to insist that it still belonged to a little nation on the other side of the world, seems to have changed the very geography.

In the heart of this most English city is the suburb of Fendalton; in its heart is Saint Barnabus Church. There I found Canon Bob Lowe, the Anglican vicar of Fendalton Parish. We drank tea. He patted tobacco into a pipe. He spoke in a deep, rumbling voice, his articulation entirely British. I could have been in any vicarage in Oxfordshire. But he spoke of separation.

"For years we've been more British than the British," he said. "Now we realize that the umbilical has been severed. Really because Mother has cleared off with a continental gentleman, and left us as orphans in the Pacific."

The continental gentleman was the European Common Market, which, when it accepted Great Britain into its ranks, ended New Zealand's favored status as supplier of much of the mother country's butter and lamb. It was a final casting off of ties that had already been weakened by New Zealand's realization in the Second World War that both her enemies and her allies ringed the Pacific, and the Atlantic was far away.

Oddly, another man with an Etonian accent, New Zealand's former ambassador to the United Nations, Tim Francis, now deputy secretary of foreign affairs, had made a similar observation back in Wellington. "This sleepy little nation," he said, "had just awakened from its European dream, looked around and discovered where it was."

"New Zealanders have suddenly become consciously aware that we are a Pacific people," he said, "and that this is going to be our home for all eternity."

In the morning I headed for the highlands. It was a gray day on the Canterbury Plains, gray as the beach at Kaikoura, but as I neared the mountains the light broke through and painted the landscape anew, as if it had suddenly become spring. Green and brown fields leaped into color; the hills developed intricate folds and canyons; the turquoise rivers shone like the precious liquid they were. Exhilarated, I pulled back on the wheel and soared up past the scattering clouds into the bright world above.

I set a course for Mount Cook, a three-cornered monolith that tops the Southern Alps at 12,349 feet. I floated westward like a barge on a river of cloud. Slowly the river dwindled away, casting me off to fly alone. Below me, the broad fields of the Canterbury Plains, with their neat seams and humid farming haze, had been replaced by an enormous, untamed land of empty brown valleys edged with bare rock and snow. Here and there enormous milky blue lakes lay against the rock. The snow was hard to look at. The air was without stain; I could see a hundred miles both north and south. More mountains.

I landed at a town called Tekapo, at the foot of a long, milky lake. There I was met by Graeme Murray, whose striking leanness was exceeded only by his enormous reservoir of good will. He is a partner in a company called Air Safaris, which flies tourists around the glaciers.

Tekapo is a small town. Its limited economy is based on tourism and the hydropower plant at the outlet of the cold lake. Nearby, a military base lies against a hill. The New Zealand armed forces were conducting maneuvers above the base. I had been advised not to fly over, lest I encounter a stray round and end my journey early.

The next morning, the air was placid. It is not always that way in these highlands. The government required that Air Safaris build its hangar to withstand the combined force of six feet of snow and 150-mile-an-hour winds. "We haven't had that much snow yet," said Richard Rayward, Murray's partner.

Rayward is the chief pilot; he got into flying back in 1964 by carving airstrips out of the dense bush of southwest South Island and flying in there to hunt deer in the wilderness. He once landed a single-engine Cessna 180 on a remote runway little longer than a football field. The plane was loaded with the equipment he needed to build the strip long enough to take off again.

I flew west, across Lake Tekapo and the huge gravel washes between it and the glaciers, over the ridge of the Alps. The flight did not take long. Broken clouds began at the west edge of the passes; I flew out into the quiet space above them. Again the river of cloud carried me west, and again it changed the world. Less than 45 minutes after leaving Tekapo, I descended and descended and descended, and finally came down into a hazy, cool, wet land, where all harsh lines of topography were smothered in foliage.

I left the plane parked on the wet grass at the airport of Fox Glacier and squelched through grass and mud over to the road. This was the West Coast, a place that people elsewhere in New Zealand speak of with a mixture of ridicule and awe. I had heard that if you live on the West Coast—at Greymouth, Hokitika, Fox Glacier, or Haast—and you buy a fine pair of leather boots, you should oil them every day and store them off the ground, and with luck they will last you a full two weeks. The average rainfall at Hokitika, about 50 miles north of Fox Glacier, is almost 120 inches a year.

Deep in its moist haze, the West Coast is fathomless and strange. The forests of *rimu, matai* and *rata* trees, and the tree ferns called *punga* crowd the roads and shut off light. You can become hopelessly lost 100 yards from home. Waterfalls drop out of fog onto treetops, misty as the rain. The bush is alive with dark green parrots and blue wood pigeons, but right there in the heart of it lie rivers of ice, grinding down from the mountains.

I was met by Doug Hendrie and his wife, Jan. Doug had been described by a friend as a hard case. "That's affectionate?" I had asked. "Oh, yes. We all like hard cases." People seem prone to nicknames here. Hendrie was "Schoolie," the town's single schoolteacher. He has the equivalent of eight grades in two rooms. "If your brain slows down, the kids are ahead of you. But by jinkers, we have some fun!" We later met John the Bar down at the hotel, and we discussed Nautical Ned and a minister, who owns a small aircraft and has thus become the Flying Nun. Hendrie had quick eyes in an angular face; his demeanor was wiry. His wife's face had softer outlines, and her eyes were as dark as the forest. She was a genuine Coaster, she said, having spent more than seven years of her life on this side of the island.

Sun shone in patches when I arrived in Fox Glacier. During the three hours I was there, the clouds assembled around the mountains and slowly pushed the sunshine out to sea. We raced around on the gravel roads in a 1967 Australian Holden automobile. We stopped right in the middle of the road when a friend driving a moving van came along. The sign on the van said "Kiwi Removals." Doug and the driver talked. "He's the hardest case," Jan said as we drove off.

We watched the tourists—who are called loopies ostensibly because they drive a big loop from Christchurch to the West Coast and back—as they walked the trail up to the foot of the glacier. Doug noted with delight that a Kea parrot was standing near the loopies' car.

"Cheeky birds," he said. "Like to peck out the rubber around the windscreen. The loopies drive away, and here's the bloody windscreen in their lap." We paused at a huge gravel wash where a recent flood had shaved all the superstructure from a bridge, leaving a battered concrete slab. The wash now contained a trickle.

"The river looks so bloody innocent," Hendrie said with a grin of satisfaction. "You should see it after a bloody 30 inches of rain." He paused, perhaps looking down into the riffle for a trace of the gold that had brought California forty-niners to the South Island in boatloads back in 1862. There was a big, rusty gold pan in the back of the Holden. But nothing shone in the gravel, in the graywacke shingle. He continued, "The last storm filled and overflowed the rain gauge at the school three times in 24 hours. Six inches each time. But then, of course, it didn't rain again for two days."

By the time I was ready to go, the clouds were low and the air was thick with drizzle. The sunlight had been exiled so far to sea that it had vanished. I jumped in the plane, chased the light, found an opening and climbed. In 25 minutes, I was back across the ridge. The air was clear, the land was parched. I landed on an airstrip that was part of an immense barren field. The West Coast was just a dream of ferns.

"There were no trees here," James Innes said. "Every single tree was introduced." I had landed at Halden Station, just 45 minutes out of Fox Glacier. Now we were taking a look at a small part of the station's 35,000 acres. James was the third generation of Inneses on this land. He was a big, lumpy man, 34 years old, rather treeless himself. No frills. No windbreaks. After talking to him on the telephone the previous day, I had concluded that he was one of those who believe words are a weakness. But now he was talkative enough, proud to discuss his innovations.

Sheep are the traditional New Zealand livestock. Innes has plenty of sheep, but Halden Station makes most of its money on deer farming—a hot new business in New Zealand. Although helicopters—including Innes's Hughes 500—still pursue hinds through the bush, hunters now catch the animals alive and sell them to farmers as brood stock. Deer are efficient manufacturers of protein, and—far more importantly—you can sell their horns to the Koreans and their flesh to the Germans for princely sums.

Deer sales are part of an odd trend in this country: luxury marketing. Kiwifruit, that delectable little fruit that looks like a dirty tennis ball but is more expensive, made millionaires out of 14-acre farmers in the 1970s. Other horticulturalists are going into avocadoes, persimmon and cherimoya, hoping for similar crazy miracles. Sheep farmers are selling the old ewes, building new six-foot fences, and starting to breed deer for the game-food restaurants. "We could afford to diversify our products into virtually catering for the up market of the world," a real estate agent told me enthusiastically. But for a nation that claims to embrace equality to turn itself into the breadbasket of the rich seems an ironic choice.

New Zealand, a country that looks so unspoiled, has, in fact, been molded by man. In the dry lands, all the trees you see were planted. In the wet lands, all the meadows you see have been cleared. This gives some farmers a deepened sense of their responsibility to the land, both for its productivity and its beauty. One man told me he was going to plant deciduous trees on a knoll for only one reason: "We need some color on the farm."

But the changes have not all been benign. The weeds are European. The fiercest creatures of the bush are Captain Cookers—wild hogs descended from those left here by that explorer. Imported rabbits once almost destroyed the grasslands. Innes poisons them every three years.

New Zealanders are so accustomed to adjusting the environment that some are now planning to pester the introduced rabbits to death with introduced fleas.

I flew east out of Halden Station. The Waitaki River ran through three man-made hydropower lakes, leaving a little of its milky glacial flour in each, then spread into gravel braids and a big curved delta near the beach. With each mile, the land became more green. Climates here are specific to each location: one 1,500-acre farm I visited got 26 inches of rain a year at one end of the acreage and 17 at the other.

One nice thing about flying around islands is that when the weather turns bad you can usually sneak down to the coast and scoot over the water. As I continued south, the overcast lowered, and I passed Dunedin at 800 feet, giving wide berth to the albatross sanctuary off Taiaroa Head. That seemed a remote and lonely place indeed, but I saw an enormous tour bus negotiating what looked like a pack trail along its windy slopes. The loopies are everywhere.

Farther south, the beaches were a sandy color, no longer gray. The green grew upon the land like a blush. By the time I began to turn the corner of South Island, heading for Invercargill, the land was a somber emerald shade under the clouds. The little rivers that ran through these fields were the color of dark beer. The next day I went looking for one of those rivers and found myself instead in a tiny museum in the village of Waikawa, on Porpoise Bay. The wind blew clean off the sea, and it seemed as if the nearest life out there was Leningradskaya Base in Antarctica. Forty-five degrees, 35 minutes south. This was the farthest away from everywhere I would get.

But the world had brought its fist down even here. In the museum, there with the ladies' sidesaddle, the party invitation printed on a sliver of *rimu* wood, the little velvet heart with its inscription "Dinna Forget," was the death notice: "Rifleman John Shankland, Jnr, killed in action in France, 29th March, 1918, in his 26th year." Below it, without comment, was another. "John Shankland, Snr, Monday, 17th June, 1918."

I spent a weekend at Waikawa and at Invercargill. "Inver" in Gaelic means the mouth of. So the name of this town really means the mouth of (William) Cargill, who was a South Island pioneer. The city is built beside a wide estuary. I began my visit to Invercargill with the true New Zealand attitude of internationalism, watching a television broadcast of a softball game between Japan and the United States being played in Christchurch. The star batter, the announcer said, was "a delightful chap." I ended it with a tour of a freezing works, the place where sheep and cattle are turned into frozen lamb, mutton and beef. The place was strikingly clean, and it was an educational visit.

I flew right up the center of the island, both seeing and missing great beauty. To the west, Fiordland National Park, a maze of canyons filled with sea, was buried in layers of wind-smoothed gray cloud. To the north I passed near Jane Peak in the Eyre Mountains and looked down upon the shining city of Queenstown, a summer and ski resort on the shore of the magnificent Lake Waikatipu. To the east only a few miles was Milford Sound, which is the most photographed place in New Zealand, but it, too, was shrouded by mist—or by the smokes of the Ngatimamoe.

On north I flew, past Mount Aspiring, past Mount Cook, through the vast dry lands of Lake Pukaki, Tekapo and Halden Station. I kept flying. There were the plains of Canterbury, and north of them a huge forest of Monterey pine trees, called *Pinus radiata* here, planted decades ago. The Kaikoura Ranges stood without cloud today, but I passed to their west, across the endless jumbled mountains of the north end of the Alps. The peaks were bare, scoured by wind and snow. At 9,500 feet I had the illusion that the entire South Island was spread out beneath me—blue coast to the east, gray coast to the west, and, to north and south, range after range of dark shapes and shadows with an occasional glint of water.

Then there were diminishing hills, little farms, windbreaks of poplar, a town. It was wonderful, coming down here from that wild region of rock, snow and geologic rawness. It was a setting of peace.

The idea of paradise is always changing. In today's world Utopia may just mean survival. On the ground at Nelson, a city of flowers and sunshine, I spoke with a slender man who had the flat light of what some people call realism in his eyes. Eyes like his are burdened by expectations that sleep does not ease. He had left the United States 19 years ago because he and his family anticipated nuclear war. He did not want me to identify him, because the last time his name was published he was deluged by requests for help from others who wished to become refugees.

The idea that this place is so far out of reach of world anger that it can escape is appealing. The smell of the summer air in New Zealand makes you homesick for places you can't remember. Clover, hay, wind off the sea. Some think the time will come when this remote nation of islands will carry the last fragrant memories of the world we know today into a charred future. When the bombs fall, just this corner of the world will remain. Then, in some distant generation, visitors may come—perhaps by canoe from Samoa—and find what you find today: the scent of something recollected, or maybe only dreamed, that you suspect you once loved.

Power. How different from mana is this kind! In order to keep that power at arms length until it goes off, people like this man are demanding a nuclear free zone in the South Pacific—no weapons, no power plants, no early warning stations, no granting of ports for nuclear vessels.

So the man in Nelson does not feel comfortable, but he has taken his precautions. He is protecting his family with what looks like distance. As it was for John Shankland, Snr., who brought up the son who was his whole life in a verdant land as far away as possible from the petty disagreements of Europe!

I left South Island in sun and a good, strong westerly. As I climbed out of Nelson, I could see little white figures playing cricket on a pitch. I crossed Mahau Sound, and Queen Charlotte Sound, and looked out on the great long beach of Cloud Bay. Then South Island was gone.

My visit in Wellington was brief and—well, breezy. Then it was north again. I was under time pressure. The next day the nation of New Zealand was to celebrate its founding—the day Captain Hobson sat down with the chiefs of the Maoris and signed the Treaty of Waitangi, the document that gave the British a new colony. But on walls in Wellington were new graffiti: "The treaty is a fraud." The statue of Queen Victoria near the Basin Reserve had been splashed with red paint. She looked wounded to the heart.

White New Zealanders think of themselves as *pakehas*, a Maori word that has been accepted into English, like mana, without negative connotations. But the terms that describe New Zealanders' racial relationships, also like the word mana, are not well defined. It is clear only that the nature of the link or the schism between the Maoris and the pakehas is at the heart of the New Zealand way of life.

For years New Zealanders have apparently thought of their nation as amazingly integrated; but now there are voices raised. Some think the new strife is honesty breaking through at last; some think the unrest is a delayed mimicking of the civil rights and black power troubles in the United States. Aside from racism, the turbulence between Maori and pakeha is often blamed either on the recent drift of Maoris from the country to the city, where the wildest of the young join motorcycle gangs, or the growing influx of other Polynesians, like the Samoans.

Whatever the cause, the focus of dispute is the Treaty of Waitangi, that document created and signed by Maori chiefs and an English sea captain in 1840. To the British, the treaty, like many New Zealand plans, was to be a model for the world: native people were not merely to be murdered and shoved aside. In the treaty the Great Queen, "regarding with Her Royal Favour the Native Chiefs and Tribes of New Zealand," gave them the rights of British citizens and promised them the "undisturbed possession of their Lands." Unless, of course, they wanted to sell.

"Beneath all the friendship between Maoris and settlers," wrote Keith Sinclair, the historian, "there lay the stubborn fact that they were rivals for the possession of the land . . . . To the settlers, land was money; but to the Maoris, it was life itself and more." Yet the Maoris sold the land, hectare by hectare, mile by mile, perhaps because, like the American Indians, they did not understand how a man could think of owning this thing that was like air and sea, free to all. By the time they learned about fences and trespassing and then began to fight, it was too late. The pakeha was here to stay.

I flew north, across fields drying out from the drought, across tiny orchards of kiwifruit or oranges, where windbreaks are planted every 30 meters to keep out the pestering wind, across little forested canyons in which the tree ferns, seen from above looked like a sprinkling of green stars. Here, in the tropical north, so unlike the broad, rugged country of South Island, New Zealand was just an arm of land reaching up into the Pacific. The haze was blue, and from 3,000 feet I could see the sea wash on either shore.

At two o'clock I landed at Keri Keri, on grass. Three hours later I watched the celebration at Waitangi. Unfortunately, it was symbolic not of independence but of dominion. The VIPs sat in chairs in front of the treaty house, where a beautiful native *pohutukawa* tree cast its shade. The public sat or stood on the grass. The police, who wore high helmets but no guns, faced them. Two years ago, when all of New Zealand erupted over the tour of the country by the South African rugby team, protestors went down to the stadiums to stare into the faces of the police. It was called "eyeballing." Here separation was more decorous. But there was tension. The police were spaced at ten-meter intervals, and seemed to hold between them strands of charged, invisible wire.

If the celebration had a message, it was a story of conquest, a history of pakeha supremacy. The day grew colder. The police put on dark blue jackets, the spectators pulled on their coats, the Maoris in their native costumes suffered. They stirred and huddled and rubbed the goosebumps on their arms. "Irrespective of how far we might be beaten," one said, "irrespective, ladies and gentlemen, we are not conquered." But they were very cold.

When I left Waitangi, I flew south then east, across 30 miles of rough sea, to Great Barrier Island. Here the *pohutukawa* trees grapple with stone and wind and survive, and the native *manuka* scrub is taking back the hillsides that pakehas once struggled to clear for sheep. People come here to live the Maoris did, treasuring the land and simplicity.

When a New Zealander has had enough of Auckland, he gets out and goes as deep into the country as he can. For some, that means the Barrier. From Auckland they come, to purchase ten acres or more on an island where there is no community power system and only a half a dozen phone lines to the rest of the world.

I stayed at the Pohutukawa Lodge, a little informal hostel and hotel. Frances Greaves cooked meals on a stove that burned *manuka* logs; Richard Greaves collected *manuka* blossom honey from hives all over the island. The bees are of European descent, toughened on New Zealand wind. "The Barrier bee," Greaves said, "is an aggressive little sod." The hotel's generator runs from seven to ten each evening. The rest of the time there's a precarious 24-volt battery system; when I turned on the light in the bathroom one night, Greaves abruptly appeared bearing a candle.

Late that night I blew out the candle and stood at the window. For a few minutes, headlights followed the road as islanders drove home from an evening's revelry at the community hall, then even that ceased. There is no darkness as profound, or as safe, as nights where there are no floodlights to make shadow. The island was at peace. I didn't want this day on Great Barrier Island to end. I stared up at the night sky. The stars seemed to grow brighter. The surf crashed on the beach. The west wind roared in the *pohutukawa* trees.

As I approached Auckland on my last flight, with 2,500 miles behind me, New Zealand's long white cloud lay overhead. Across the harbor I flew, descending at great speed past Lone Tree Hill. As if preparing you for the accelerated pace of life in this city, the rules at Auckland International require all aircraft to maintain 90 knots almost to touchdown; that was about as fast as this plane could go. On final, lined up for the runway, I came charging in, head down, engine roaring, as if I was glad to get it over. As if I was glad to get it over?

There is so much to remember in the last few hundred feet, while the engine noise dwindles away in the glide, and the air is finally calm. I recalled, of all things, drinking tea in the middle of a forest of native *totora, rimu, matai* and *kaihikatea* mixed amongst imported *Pinus radiata* and Douglas fir. I sat in a lookout with a little Kiwi woman with bright pink English cheeks. We were surrounded by windows. From horizon to horizon the great trees spread—the tawny natives and the deep green newcomers—their roots deep together in the pumice soil of *Aotearoa*. [1983]

Molokai:
Where Old Hawaii Lingers

WILLIAM ECENBARGER

IT'S FIFTY-THREE MILES and 100 years from Honolulu to Molokai. Nearly everyone aboard the 20-minute Hawaiian Air flight is heading home after a day of shopping in Honolulu. They carry plastic bags with department store logos and takeouts from Pizza Hut. Children, precariously perched at the ends of their parents' patience, smuggle giggles to each other.

From the air Molokai looks like a slipper, 38 miles from heel to toe. Just now it is tucked under a blanket of quality clouds that cover the island's two principal landmasses—each formed by a volcano.

From the eastern tip of Molokai—fifth largest of the 132 islands, exposed reefs, and shoals that make up Hawaii—proud residents can look across nine miles of open water and see what happened to the once pristine beaches of Maui. Development has a foothold here, but of the estimated 5 million tourists who went to Hawaii last year, only about 15,000 spent a night on Molokai. Most of the island's tourist attractions have a unique homegrown flavor.

Hawaiian Air makes a smooth-as-silk landing at Hoolehua Airport, and the shoppers move through the terminal quickly with their packages and still warm pizzas. There are no metal detectors or X-ray machines in sight. The terminal clock is 20 minutes slow. You peek into a tiny bar, half expecting to see Bogey and Bacall having a drink.

There are no traffic lights, fast food emporiums, or movie theaters, and no building is taller than a coconut tree. About 6,500 people live here. Four thousand of them are of Hawaiian ancestry—the highest percentage on any major island—and more than a third of those speak little English.

Molokai is a relatively unspoiled slice of Polynesia, a half-forgotten remnant of old Hawaii, where people deal with each other as though they will meet again, where the days are so congenial you want to invite them back, and where birth and death are important events. Life chases its own tail.

It is a place where the outsider needs to tiptoe and hold down the noise. A place Jack London would have liked and written about—and did. A place to be sniffed, rolled on the tongue, and drunk slowly. Molokai leaves a wake, and you bob around in it long after you've departed.

EIGHT MILES SOUTHEAST OF THE AIRPORT LIES MOLOKAI'S PRINCIPAL TOWN, Kaunakakai, whose name became famous in the 1920s with a popular song called "The Cockeyed Mayor of Kaunakakai." There is, in fact, no mayor of Kaunakakai because all of Molokai is part of Maui County, which has a mayor at Wailuku on the island of Maui.

Downtown Kaunakakai consists of three blocks of false-front stores along a street called Ala Malama. Except for the palm trees, it could be a town out of frontier Arizona. Most of the islanders come here to do their basic shopping on weekends, and this morning two women are having a cart-to-cart talk outside Imamura's dry goods store, their voices rubbing cheerfully against each other. The mahogany odor of coffee wafts out of the Kanemitsu Bakery, which has been making its famous Molokai bread for 61 years. Men in shorts, sandals, T-shirts, and baseball caps sip and wag old tales. Their words tumble over each other as coal down a chute.

The Mid Nite Inn, which closes promptly at 9 P.M. every night, offers the catch of local fishermen—*aku* (skipjack tuna), *kākū* (barracuda), and *ōpakapaka* (pink snapper)—grilled on a diesel stove. Served with rice and kimchi, Korean-style cabbage, the fish dinners cost about six dollars. The restaurant got its name a half century ago when departing travelers would come here for a late dinner while waiting for the interisland ferry to leave at midnight.

Nearby a bulletin board asks, "Should Molokai Be a Playground for the Rich?" and underneath someone has tacked up a typed paragraph: "Forget this 'Molokai, the Friendly Island' crap. That's Mickey Mouse stuff. Why not call it 'Molokai, the Savage Island?' Make it hard to get to. First tourist off the plane every morning is notified he won first prize: He will be the human sacrifice at nightfall. Get some guy to go off in the hills and beat on drums all night. Have some big Hawaiian walk through the lobby in a jockstrap every afternoon. And when someone complains or asks who he is, you ask, 'Who?' You didn't see anything. Kill the pig for the luau on the porch at lunchtime so everyone can hear the screams and see the blood."

Much of Molokai's populace is standing up against outside developers, setting up a dike against the wave of fashionable thinking that has carried away so much of the world's natural beauty. In 1980 a grass-roots group successfully opposed plans by a California developer to build luxury condominiums near the old Hawaiian settlement of Pukoo. In 1986 the Swig family, operators of San Francisco's Fairmont Hotel, pulled away from plans to build a large hotel on Puu O Kaiaka, a rock that is held sacred by some Hawaiians. But in 1988 Del Monte Foods closed Molokai's last pineapple plantation, and at one point the island's unemployment rate rose to about 20 percent. How long Molokai can fight development under these conditions is uncertain. Money not only talks, it keeps up a running conversation.

Down at Kaunakakai's wharf, the huge barges are gone that used to loan tons of Molokai pineapple for shipment to a Honolulu cannery. A ferry called the *Maui Princess* is pulling out to sea, carrying more than 100 Molokai residents to the glitzy resorts of Maui's Kaanapali Beach, where they will serve meals and change linens in the tourist resorts.

DEVELOPMENT HAS ESTABLISHED A BEACHHEAD ON MOLOKAI'S WEST END, where the roar of the surf is punctuated by the purr of golf carts and the *pock-pock-pock* of tennis matches. The 292-room Kaluakoi Hotel and Golf Club, which began life 12 years ago as the Sheraton Molokai, was purchased in 1987 by a Japanese real estate operation for $35 million. The resort sits majestically overlooking an almost deserted stretch of Kepuhi Beach, and at night one can see the lights of Honolulu carelessly splashing kilowatts into the sky, providing an encore to the sunset. Everything here is first-class, but Kaluakoi has been a losing proposition thus far.

In the center of the island the red earth around Kualapuu is rinsed and puddled by a recent rain. Machinery sits idle and rusting in the Del Monte pineapple fields, and a spider is mending a broken windowpane in the plantation office. Cotton was planted on Molokai during the Civil War to supply the North, but it didn't work out. An attempt to grow sugar was abandoned in 1898 because the irrigation water was too salty. During the early part of the nineteenth century, Molokai was the world's leading producer of honey, but an epidemic struck the hives in 1957, and the industry collapsed. In 1986 the government slaughtered all the island's cattle in an effort to wipe out bovine tuberculosis.

North of Kualapuu, one industry does seem to be thriving. There are fields of what appear to be tiny A-frame dollhouses. In fact, they are the homes of Molokai's "feathered gladiators"—fighting cocks that have been raised on the island for as long as anyone can remember. Cockfights are very illegal but very popular on the island.

Macadamias are the specialty at Purdy's Nuts, which operates a stand not far from the airport. The air is redolent with burning leaves, but Theo Purdy, mother of owner Tuddie Purdy, apologizes for the smoke and explains that brush must be burned every day because the macadamias are harvested from the ground. She also apologizes for the profusion of cats, which are there to control the rats that would otherwise climb the trees at night and eat the nuts. Then her eyes crinkle with laughter, and she slices a fresh pineapple and serves it on a plate-size candlenut leaf. Tuddie arrives, runs his sleeve across a brow wet with honest sweat, cracks a coconut, slices it, and places it on a papaya leaf. He says macadamia trees were brought to Hawaii from Australia in the late 1800s as he offers a half-coconut shell full of roasted macadamias. Theo says the Purdy family has run the nut farm—a small grove originally planted in the 1920s—for eight years, and puts out a jar of macadamia blossom honey. There are two other nut groves on the island. Like Purdy's, they are small and family operated.

THE STATE HIGHWAY KEEPS GOING BEYOND THE TURNOFF FOR THE KALUAKOI resort, but after about a mile it seems to think better of the idea and just stops. At the end of the asphalt one finds Maunaloa town, which was built by the Libby interests in 1923 with prefabricated houses shipped from the mainland. The enclave ceased being a company town in 1975, when the last pineapples were picked, but for several years now it has been one of the world's most laid-back shopping districts. Enterprises include the Big Wind Kite Factory, Plantation Gallery, Dolly Hale Handcrafted Dolls, JoJo's Cafe, Red Dirt Shirts, the Maunaloa General Store, and those of several artists.

Bill Decker, a woodcarver, extends a tourniquet handshake, tells you how he's planning a trip around the world, tells you how he wants to write a book about the world's great woodcarvers, and tells you how he came here because Montana was getting too crowded. He runs a huge hand through his hair, which is long and blond, and makes a path with a sandaled foot through a six-inch clutter of wood shavings, sawdust, Marlboro butts, and beer cans. Then he throws another monologue in the fire.

"A lot of bartering goes on around here. I traded one of the first signs I made for a 1967 Ford Falcon, and later I traded another sign for a truck. The car and the truck fell apart, but my signs are still there. Signs are my bread and butter. The people are really friendly all over Molokai. Somebody cuts down a tree, they call me right away."

Over at the Big Wind Kite Factory, mainlander Paige Rodrigues says she came here with her Honolulu-born husband, Antone, about three years ago. "We eloped with my two young children, lived on the beach for three weeks, and finally came here. We started making kites to keep the kids busy, and before we knew it we were in the business. We work hard making kites. I wouldn't want to see Molokai all built up and ugly." She shoos one of her 14 cats out the door. "I love it here. A lot of people who have lived here all their lives just don't know what they have."

At the Maunaloa Post Office, Joy Kaupu, who has lived here all her life, hands an old man his mail with a smile you could pour onto a waffle. She's the part-time postal clerk and substitute teacher at the 100-pupil elementary school. "It used to be really beautiful here, but now it's not so nice. People come and go. There are no jobs. We have 180 boxes, but a lot of them are empty."

Next door groups of men and women sit under a tin roof. The women gesture and talk at 100 words a minute, with gusts of up to 150. Some men are playing cards, periodically throwing money into the middle of the table. Others just sit and smoke, like shipwrecks on a reef.

THE WORLD'S HIGHEST SEA CLIFFS, SOME 3,000 FEET TALL, ARCH THEIR BACKS against the relentless waves on the north coast of Molokai. From here the Kalaupapa Peninsula juts out in the Pacific, low and flat, like the fork on a forklift. Kalaupapa is sealed off from the rest of the island by a 1,600-foot, nearly perpendicular cliff, a *pali*, and it is sealed off from the rest of the world by crashing surf on three sides. Geography tugs at history, and Kalaupapa's isolation made it the site of one of the most infamous of all leper colonies. Native Hawaiians had little immunity to the communicable diseases that were introduced with the arrival of outsiders. And Hansen's disease—leprosy—reached an epidemic level in the 19th century, affecting one in every 50 people on the islands. Public panic grew, and in 1865 King Kamehameha decreed that all lepers should be banished to Molokai. They were dumped from ships and left with no medical treatment or shelter and only meager food.

The arrival of a Belgian priest known as Father Damien in 1873 changed the course of the settlement, and new medical knowledge brought the disease under control in the 1940s. Today Kalaupapa is home to 93 former patients who could leave, but who have chosen to live out their days here.

If you're in a hurry, you can get a seven-minute flight from Hoolehua Airport to Kalaupapa's dragstrip runway. But the once-in-a-lifetime way to go is by mule down the cliff on a narrow, rocky, 3-mile trail with 26 switchbacks. For many years this was the main route in and out of the leper settlement. Now most supplies come in by air, and twice a year, when the seas are calm, barges bring in large items like cars and furniture.

The mule trail, which was carved out in 1886, runs through a rain forest lush with mangoes and guavas, and only the muleskinners' portable radio, offering John Denver's "Thank God I'm a Country Boy," keeps the rider in the twentieth century. You look down and see only your own denimed leg and then the Pacific some 1,500 feet below. When Jack London rode a mule into Kalaupapa in 1907, he noted that a single misstep meant a half-mile drop "through the blue space into the blue ocean." At first you hear the rolling surf as background to the *clop-clop* of the mules' hooves. But the ocean turns up the volume steadily during the descent, and finally it eclipses even the radio. The trail levels out at the bottom and hugs the coast. Ten-foot waves double their fists and pound the shore.

The mule train is greeted by Richard Marks, a former leprosy patient who lives here and operates tours of Kalaupapa in a beat-up school bus. Marks is reconnecting a battery cable that popped off as he was driving on a dirt road with a dozen tourists.

"The only way Hawaiians are going to hold on to this island is through farming," he says. "There's a lot of water here, but it's all in the valleys, and no one wants to bring it down to the farmland. Once you got the water where it would do some good"—he slams the hood of the bus closed—"Molokai could become the breadbasket of Hawaii. If the resort people get hold of the land, they'll find enough water for 5,000 hotel rooms and 26 golf courses."

Kalaupapa seems secure from development, though. It has been designated a national historical park, and for several years the National Park Service has been working to preserve and restore the peninsula. There have been no admissions to Kalaupapa since 1969, but the state has guaranteed that the remaining former patients, many of whom were brought here as children, can stay. Their average age is 68. They live in cottages in Kalaupapa, where there is a store, a gas station (open once a week), a small hospital, a bar, three churches, and a post office (zip code 96742).

MANY PEOPLE CONSIDER THE HALAWA VALLEY TO BE THE MOST BEAUTIFUL, UNspoiled area of all Hawaii. To get there, you head east out of Kaunakakai on State Route 450. Along the coast, ruins of ancient rock walls rise out of the surf. These are the remnants of royal fish ponds, built between the fifteenth and eighteenth centuries to provide a steady supply of fish delicacies for the Hawaiian *ali'i*, or chiefs. Missionary churches, their roofs pitched like praying hands, dot the landscape, and mongooses make death-defying dashes in front of your car. The Mapulehu Mango Grove—with 2,500 trees—comes up suddenly on the right.

Larry Helm, wearing a "Don't Worry—Be Happy" T-shirt, apologizes because mangoes are out of season, grabs a fresh pineapple, gives it an executioner's whack, and hands slices to his visitors. The mango patch had been abandoned for a long time, and he and two partners leased it last year and are trying to turn a profit. To supplement their income, they are offering tourists wagon rides to a nearby *heiau*—an ancient place of worship, built several hundred years ago. These tours are followed by a dinner and party on the beach, complete with net fishing, coconut husking, and authentic Hawaiian music.

"Hawaiians, especially younger Hawaiians, are beginning to take real pride in their heritage," says Helm. "We want visitors to know that there is more to Hawaii than hulas, ukuleles, and leis."

The road narrows as it climbs into the valley, and the temperature drops noticeably. There begins a dizzying unraveling of macadam, hugging cliffs, and switchbacks. One minute the emerald meadows of the Pacific are on your left, the next minute they're on your right and to your left are green mountains wrapped in robes of clouds. Just offshore is Mokuhooniki island, which looks like a huge turtle, and was a bombing practice target during World War II. Just before the highway begins its final three-mile descent, there is a grove of gray-barked candlenut trees that is the most revered spot on all of Molokai. The trees were planted by Lanikaula—the greatest of the kahunas, or Hawaiian priests—who is buried here. Many Hawaiians won't go near this grove, lest they offend its sacred nature, and when Del Monte wanted to clear the land 20 years ago, not a single local worker would fell the trees.

The highway ends in the Halawa Valley, which was a thriving community of taro farmers until 1946, when a 36-foot tsunami swept through from the coast. A few families and individuals have returned to the valley to live without electricity or telephones. One of them is Dupre Dudoit, 59, who provides parking and security for people hiking the two-mile trail to Moaula Falls—actually two successive falls with a large pool at their base. Dudoit also sells soft drinks, beer, and bananas from the porch of his trailer. He humps his shoulders like a roosted bird, says he would never want to live anywhere else, and gives a handwritten sheet with directions to the falls. The legend is that it's only safe to swim in the pool if you throw in a ti leaf and it floats. If the leaf goes to the bottom, it means the water spirits are calling for someone.

The trail leads through tropical forest. Wild ginger blossoms varnish the air. Wild avocado and guava line the way. It is everyone's South Seas fantasy—not the South Seas of the tourist brochures, but the South Seas of Jack London and Somerset Maugham. The land is throbbing, strummed, as though touched by an unseen hand. [1989]

Enchanting Chiloé

TIM CAHILL

UMILIANA CARDENAS Saldivia still doesn't know exactly what she did to get herself in trouble with the witches and wizards of Chiloé. A tiny, ebullient woman who appears to be in her early 60s and looks a bit like Dr. Ruth, the television sex therapist, she is the author of a book about *brujos*, or witches, on this large, verdant, heavily forested island off southern Chile.

Some stories in her book tell of brujos at play (one of them made a large pot walk about on its iron legs), others of brujos at work (one unfortunate Chilote came home to find a large bear in his bed—this on an island bereft of bears).

Yes, brujos could be vengeful. And somehow Umiliana had fallen afoul of one of them. Maybe several. "Brujos," she said in Spanish, "can be very evil. They cause death and illness." Evil witches and woods? A bear in the bed? Umiliana's words twanged at some of my own childhood memories.

We were sitting in a waterfront restaurant in the capital city of Castro, a town set high on a hill overlooking the ocean. Paved streets plummeted down the hill to the line of restaurants that faced the water. Ours was not the most elegant of establishments. It was something of a dive.

There were rough-looking men—fishermen and day laborers—drinking heavily at the bar at ten in the morning. They were big men, descendants of the Spanish conquerors, and of the Italian and German immigrants who came later; there was little or no visible evidence that any of their people had intermarried with local Indians.

The men were drinking *pisco*, a fiery pale brandy made from the first pressing of grapes. The most popular brand was Pisco Control, and the popular drink was the pisco sour, a blend of two-thirds pisco, one-third lemon juice, a bit of sugar, and egg white, shaken with crushed ice. Five or six of these can drop an untrained drinker to his knees. Take my word for it.

Meanwhile, a waiter wearing a red vest and black bow tie placed my breakfast on the table: grilled sea bass, baked bread still warm from the oven, and freshly squeezed orange juice. I considered the hard-working, hard-drinking men at the bar, I considered this waterfront setting with civilized service and great food, and I caught a vague hint of San Francisco at the turn of the century.

But there was something else here as well, and I tried to put my finger on it as Umiliana told me about her problems with the brujos. In 1956, when her first daughter was born, Umiliana had trouble breast-feeding, and doctors found slivers of wood under her skin. (She pounded her chest to show where they found the wood.) And then she began finding dogs and cats in the house every morning, though she locked all the doors at night. For nine full years she never felt entirely well.

Umiliana suspected brujos. She investigated and discovered that there was, in fact, an association of brujos. In their initiation rites, she learned, men and women went to Chiloé's outlying, uninhabited islands and walked naked in the frosty cold of winter, three times around the high-tide line to make a deal with the devil in exchange for power.

Umiliana said she stumbled onto a woman who was head of the association of brujos. She was, Umiliana said, a good woman, an atypical brujo, one who wanted to help people. For a certain sum of money, Umiliana was given a document insuring her health and that of her family. Since that document was signed, in 1965, neither Umiliana nor anyone in her family has had a single sick day.

There were still brujos everywhere on the island, Umiliana said. Sometimes, at night, in the deep, fog-shrouded forests, they danced their ecstatic brujo dances, reconfirming their pacts with the devil. God help anyone who stumbled onto such a gathering.

And then it came to me, the odd sense of deja vu that these stories of witches and woods and bears in the bed engendered. The Germans had come to Isla Grande de Chiloé around the turn of the century, farmers looking for land in the dense forests, men and women familiar with the works of the brothers Grimm, who had collected folk tales of the forests of Germany, Scandinavia, and the Netherlands nearly a century earlier. Here, on Chiloé, people still believed those tales, or tales very much like them, and they told the stories at night, sometimes by the light of kerosene lamps.

Chiloé. The last fairy tale island.

CHILOE, MORE THAN 150 MILES LONG, IS THE SECOND LARGEST ISLAND IN South America. (Only Tierra del Fuego is bigger.) The island lies off the Pacific coast of Chile at the 42nd degree of south latitude, about the same position south as Coos Bay, Oregon, is north. The weather is about the same in both places: cool damp summers, cold (and sometimes snowy) winters moderated by the marine environment. The sea is often shrouded in fog, but the day was clear as I took a ferry across the narrow strait from the mainland. Dolphins raced the boat, and small penguins porpoised alongside, flying through the cold water with a grace they would never exhibit on land.

Visitors to the island almost always come from mainland Chile, generally from the capital city of Santiago. Renato Arancibia and his wife Isobel, for example, once worked as travel agents in Santiago. After the births of their two boys, the couple took stock of their situation. It was a two-hour commute to and from work every day. What was the purpose of having children if you had no time to enjoy them? A few years ago the family visited Chiloé. The people were, in Renato's words, "simple," by which he meant they were men and women of the land and sea: upright, honest folk, and very shy. The island was "tranquil"—a word every Chilean tourist seems to use at least once in describing Chiloé—and a good place to raise children.

Renato and Isobel set up a travel-oriented business across from the market on the waterfront at Castro. His company, Pehuén Expediciones, rents mountain bikes and runs party boats to the small outlying islands, where the last vestiges of "old Chiloé" are to be found. It's still a hand-to-mouth business. Tourists, especially foreign tourists, haven't quite caught on yet.

I PEDALED MY RENTED BIKE PAST THE CASTRO MARKET WHERE COLORFUL, hand-knitted wool sweaters sold for about seven dollars. The road wound down the waterfront, and past the *palafitos*, homes of fishermen that extend out over the sea on great stilts made of local hardwood. The houses, shingled in weatherbeaten wood, are very picturesque. (Indeed, they are best enjoyed in pictures, because garbage and sewage are dumped into the sea from these homes, and signs warn that it is unsafe to swim.)

The palafitos are a holdover from the old days, when there were no roads on Chiloé, and merchants sold sugar and salt from boats. Today, there is a paved road bisecting the island from north to south. (OK, the last ten miles or so aren't paved, but they are negotiable by rental car.)

The road I was riding cut across a bridge and rose into the steep farmland across from Castro. The sky was several different shades of gray, and a breeze set wildflowers swaying in the fields.

Pastures, apple orchards, wheat and potato fields were set out in square patches. The farmers had left the forest intact at the periphery of their fields, and the undulating, alternating patterns of agriculture—the dark green of the potato fields, the gold of the wheat, the verdant forest surrounding them—made the island seem softly sculptured.

Robust dairy cattle shared pastures with small island horses. Beyond the fences, where the animals couldn't get at it, the land exploded in vegetation—blackberry bushes, bright red drooping flowers that looked a bit like Indian paintbrush, yellow snapdragons, and a kind of purple flowering clover. For a moment the sun bullied its way through the clouds, and several shafts of purely celestial light fell across the landscape, so that Chiloé seemed a kind of Eden, complete with birdsong.

Seven hundred feet below, I could see the ocean inlet, shaped like a piece of a particularly baroque jigsaw puzzle. Tiny figures were digging for clams near the market. The Catholic cathedral stood on a rise and fronted the town square, dominating the landscape. The church was a huge, ornate affair, paneled in tin and painted a strange, almost iridescent orange. The intricately shingled houses of Chiloé, complete with their gables and battlements, are often painted in fever-bright colors. Tradition has it that the colors help fishermen at sea locate their homes in the fog.

I scanned the water not for a boat, but for a home traversing the sea. For reasons that remain impenetrable, people on Chiloé and the outlying islands sometimes need to move their homes. The house is rolled down to the sea on logs, with dozens of men singing and whistling and pulling at ropes and driving oxen. Then—and I've seen many pictures of this—the house is floated on the sea and dragged, by boat, maybe 15 miles to another island. Only the roof protrudes above the water.

When the house is finally resituated on another island ("Honey, don't you think it would look better over here?") there is generally a *curanto*, the seafood equivalent of an American barbecue. I intended to treat myself to a curanto (the dish was advertised outside most of Castro's restaurants), but later. First I wanted to see more of the island.

TINY CUCAO, THE ONLY VILLAGE ON CHILOE'S WEST COAST, IS SET ALONG AN immense, gently curving, gray gravel beach, ten miles long and guarded by rock spires on either end.

The Pacific Ocean thundered into shore in huge breakers. Men on horseback, wearing ponchos and wide-brimmed hats, carried nets into the surf to snare sea bass. Women, bundled up in multiple skirts against the chill of the water, stood knee-deep in the foam, looking for all the world as if they were dancing to music of the ocean. In point of fact, they were digging in the sand, feeling with their bare feet for a kind of shellfish called *macha*.

Down the beach other men spurred their horses into a top-speed gallop. The horses cut sharp left, then right, agile as cats, then reared up on hind legs, holding the pose, their forelegs pawing the air. The men were practicing for the local rodeo, always held in January or February.

Most of the villages have a small stadium that looks a bit like a bullring, but a Chiloé rodeo features only one event, and a bloodless one at that. A bull is released. Pursued by a man on horseback, it is driven against the high, circular barrier fence. When the bull tires, the horse rears up and pins it to the barrier for ten full seconds. During that time, neither the horse's forelegs nor the bull's forelegs should touch the ground.

Rising above the beach at Cucao are the impossibly emerald mountains of Chiloé National Park. These coastal mountains—protected since 1982 from lumbering interests—are a chaos of erupting vegetation, so thick that the government built a wooden walkway through the woods. The forest is particularly dark: There is a twilight gloom even at high noon on a sunny day.

In the clearings, large elephant-ear plants called *malca* look fragile but feel like rough leather. A strong odor of rotting organic matter mingles with the fragrance of living things—parasitic flowers—amid a marshy land veined with small, tea-colored streams.

Trees covered with moss and lichen grow in a twisting, slow-motion lunge, looking for the best place to steal the sunlight from nearby competitors. The losers in this agonizing game of life-and-death fall but seldom reach the ground. Such is the sheer proliferation of greenery that dead trees, held by the living ones, rot aloft.

Walking along the boardwalk, I heard the warbling, half-loon, half-meadowlark call of a bird called the *chucao*, and it was coming from my right side. This, I had been told, is good luck. I decided to tempt that luck by slipping off into the forest.

The marshy ground took my boot to the ankle for the first two steps. Then I crawled, creepy-damp, through the choked underbrush until I was out of sight of the wooden walkway. One tree dominated this section of forest, a great, straight-trunked giant that rose above all the others like a monstrous stalk of broccoli. A kind of warm organic fog steamed up off a dark brown stream to one side.

It occurred to me that this was exactly the kind of forest that gave rise to the grim tales of brujos, to a mythology that included the half-human creatures I had heard about: Trauco, Machucho, and others, all of them dangerous and sinister characters, like the familiar wolves and witches of my childhood.

I am a man unmoved by superstition but when the chucao called on my left side—bad luck for sure, Chilotes say—I decided to make my way back to the walkway. I did this in some haste, and managed to scratch both arms crawling through the thorny vegetation—all the while assuring myself that I am a man unmoved by superstition.

OCTAVIO, A PLUMP, JOLLY MAN, WAS THE PROPRIETOR OF A CASTRO RESTAURANT called, not surprisingly, Octavio. He wore one of those thin, door-to-door salesman's mustaches and joined my table for a glass of Chilean wine. His establishment was a dark, bare-wood, windowless cavern that you entered through a long, unlit hallway. It was a friendly, family-run operation: Fresh-baked bread and a bottle of the best local cabernet, which was excellent, cost all of $6.

Octavio said there wasn't much of a café society on Chiloé, not during the winter months anyway. He had to make his money during the summer season, from December through February. His restaurant wasn't as fancy as some, and there was no view, so he depended on the quality of his food to draw customers. Every year, he said, the same tourists come back for more of his curanto, his grilled conger eel (it tastes a bit like halibut), his salmon, his oysters, and the seafood stew called *paila marinas*.

Originally, curanto was a kind of Chiloé survival dish. At the end of the summer, when the water got too cold for diving, fishermen collected great quantities of shellfish, dug a hole in the ground, and started a fire. After a time, the fire was covered with rocks, which in turn were covered with malca leaves. The fruits of the sea were piled on, and, over the course of hours, a thick fragrant soup developed as the shellfish were smoked for future use.

The curanto, like a barbecue, is still a social affair, and herdsmen today may add mutton or beef or sausages to the portion of the stew to be eaten on the spot.

The waiter placed an enormous bowl of curanto on the table in front of me.

"Where do they have the best food in the world?" Octavio asked me.

"Right here," I said, ever the diplomat.

"No. Tell me the place where everyone says the food is the best."

"Well, France, I suppose."

Octavio smiled brilliantly. "There are many people," he said, "who come from France. Every year. Just to eat my curanto."

I tasted Octavio's famous curanto. It confirmed every word he said.

RENATO ARANCIBIA TOLD ME THAT I SHOULD SEE THE REAL CHILOE, WHICH WAS not Chiloé itself, but its outlying islands, where there were no roads, few visitors, and no telephones. Local radio stations broadcast hourly messages: from one family to another; from a merchant with a load of goods; from a man in love to a woman waiting by her portable Panasonic.

The sea was glassy, calm, tranquil, and took on the color of the sky, which was to say, it changed throughout the day. One moment the vast expanse of water seemed lifeless and forlorn, cold and gray as iron. And then the sun burst through for an hour, and the water was cobalt blue, clear as crystal, and I could see clouds of bait fish going about their single-minded business 15 feet under the surface.

The largest settlement on any of the outlying islands is always the port. The village may consist of a dozen houses and an enormous church bigger than all of the habitations combined.

On Chelin the sound of mournful singing wafted out of the church—a funeral, perhaps—as I walked up to the cemetery set on a hillock above the village. The graves were contained in small wooden houses about eight feet long. Some of the houses were paneled with shingles. Inside, rusting paint pots held fresh daisies, and crosses were set above small altars on which pictures of the deceased had been placed. *Our Dear Mother, Catalina Santana, Rest in Peace.* The graveyard itself was overgrown. One or two of the small houses had fallen into disrepair and were filled with an explosion of wildflowers and ferns. The trunks of small trees snaked out of broken windows.

On the island of Mechuque, I saw a man building a 45-foot fishing boat, working without plans, hammering out the graceful, swooping lines that would identify the craft as his work. His family had been the boat builders of Mechuque for generations.

There were palafitos built along a riverbed that drained and filled with the tide. A man stumbled out of the forest with a load of firewood on his back while children whooped and squealed at play on the beach.

Later, I stopped to talk with Don Paulino, a gentleman of 86 years, who lived alone in an old wooden house with great, high ceilings and dull green walls. In the parlor hung an old black-and-white photo that had been hand colored so that a golden light haloed the face of the determined-looking young woman who had been Don Paulino's wife. A large horsefly buzzed loudly in the silence.

Don Paulino had been born on Chiloé, but had left at the age of 14.

In those days there was no school on these islands. Chiloé had been the last refuge of the Spanish, the last royal foothold in Chile, and the government ignored the needs of the remote islands.

Don Paulino had traveled to Argentina to work the sheep ranches and noticed that rich men owned land. When he finally returned to Chiloé, he bought land of his own, then worked at sea to earn more money. Eventually he owned two cargo boats—framed black-and-white photos of the two ships also hung on the wall—and he put the money from those ships back into the land.

Now, Don Paulino said, he was a very rich man. But it was all on paper, in deeds. He needed cash. This man—who never went to school, who taught himself to read and write—had many grandsons and granddaughters he wanted to send to college. He was now selling his land and timber to the Japanese, he said, for the sake of his grandchildren.

THERE WERE NO RESTAURANTS ON MECHUQUE, BUT I WAS TOLD THAT A WOMAN named Dina Paillocar provided a good lunch at low cost. Her simple wooden house was clean and bright. A picture torn neatly from a magazine was tacked on the wall at eye level. It showed a friendly looking lion with large blue eyes, captioned in Spanish: "Today is a marvelous day." Dina served marinated raw clams, a soup of rice and smoked fish, followed by fried clam cakes. Everything was delicious, and Dina was effervescent, indomitable. She talked about her husband who had gone to work in Punta Arenas. He had been gone a long time, she said, when she heard that he had been seen with another woman. Dina went to Punta Arenas, confronted her husband, and told him that he could have one more night with the hussy, then had to come back to Mechuque and help her raise their son. She never saw him again.

"The one more night," Dina said, "it wasn't a good idea, I guess."

"Probably not," I agreed, and Dina's quiet laughter flowed like a stream in summer.

She was, she said, better off without her husband. She had loved to cook and earned some money doing it. Her son was 14, very smart, and would earn a scholarship.

She also said she had survived many harrowing experiences in the forest behind her house. Once, she even saw Trauco.

"Oh?"

"Yes. He was a little man, perhaps three feet tall. He wore a cloak of moss and a pointed hat made of lichen."

She had been out cutting wood in the forest when she heard the sound of an ax. One single whack, and then the sound of a falling tree. Trauco: the man who could fell a tree in one stroke. Dina turned and fled. Trauco can kill with a look, she said. He can bring sickness on a bad wind. He often makes young girls pregnant.

"Trauco makes young girls pregnant?" I asked.

"Oh yes," Dina said brightly, "many parents here have sued Trauco. In court. Because he had made their daughters pregnant."

We drank some wine, and she told me about Machucho, the man with three legs, who can jump 50 feet at a bound, each step sounding like the booming of a great cannon. And about Caleuche, a ghost ship that appears in the fog. Loud music and laughter echoes aboard this ship (which you sometimes can hear in the fog). When fishermen fail to return from the sea, it is assumed they have joined the party on Caleuche.

Sun poured in through the window. The wine, a Rhine from a vineyard outside of Santiago, was tartly crisp. The day, I realized, was just as advertised. It *was* a marvelous day, on a fairy tale island. [1991]

On the Edge of Life in the South Shetland Islands

MICHAEL PARFIT

IN THE HARBOR AT DECEPTION Island, which is not as safe as it looks, someone has painted faces on five huge old fuel storage tanks that stand listing in volcanic dust and gravel. There is one face for each tank, twenty feet tall. They stare out across the bay of Port Foster, permanently astonished at the penguins that come to strut beside them, at the fur seals that bark on the shore, and at the growing number of odd ships that steam slowly through the clouds of cape pigeons at Neptune's Bellows. The ships bear people themselves astonished to be at last in Antarctica, who come to stare back at the faces and the weird birds.

The South Shetland Islands, which lie just north of the Antarctic Peninsula, are the most accessible part of this remote, dramatic region of the earth. Deception Island has been occupied since the early nineteenth century, but the presence of people on these cold and stony shores has always been precarious. Even today, with new military and scientific bases popping up all over the islands, with nations sending colonists, and with tourists and adventurers pouring south in ships, planes, and even rowboats, these islands have not become suddenly gentle. They are still on the hard edge of life.

I have been lucky enough to visit the Antarctic Peninsula area three times, in summer, fall, and winter. I will go again when I can, because there is no forgetting the place. These islands, like the faces on the tanks, haunt one's civilized dreams. Frank Wild, who went again and again to

Antarctica in the early years of the twentieth century, and was marooned there for 105 days, said that you returned because you could never escape "the little voices." What do the little voices whisper? They remind you of all the faces you have seen in Antarctica: faces on tanks, faces of friends, the many contrary, wonderful, and deceiving faces of the island.

Although tourist flights to King George Island are becoming more common, most people still get to Antarctica by ship, sailing south from Punta Arenas or Ushuaia, past Cape Horn and across the Drake Passage. The Drake is usually a ferocious place, but the three- or four-day passage through the maelstrom can sometimes show a different profile. I crossed it once in late April, expecting autumn violence, and found it so gentle that it deserved an alternative name: Lake Drake. The islands of the Antarctic Peninsula are equally unpredictable. They're exposed to the lash of the Drake, but also to the tempering effect of the sea. While the rest of the Antarctic coast is locked away for the winter by the ice pack, the islands are often accessible to ships year-round.

IT IS SAID THAT THE BEST MOMENTS IN ANTARCTIC TRAVEL ARE WHEN YOU SEE your first iceberg and when you see your last iceberg. Antarctic territory officially begins at 60 degrees south, but a ship approaching Antarctica usually passes its first big chunk of ice somewhere after Smith Island's peak has risen like a pale moon 50 miles away over the horizon. Beautiful, deadly, the huge piece of ice wallows slowly in the swell, a piece of the southern landscape broken off, drifting north to show Antarctica's first face to arriving strangers. It is massive and delicate, softly blue, but up on the bridge the crew is looking out for the pieces it launches like torpedoes, the chunks of ice called growlers that lie awash and invisible to the radar and yet are big enough to cripple a ship.

Depending on the course and the schedule, some ships reach the shelter of the South Shetlands by turning around the northern tip of King George. Others slip between Smith Island and Snow Island, a mountain to starboard and an ice field to port. But somewhere along the journey almost everyone passes through Neptune's Bellows into Deception, so that is where my own little voices take me first.

In 1820 Nathaniel Palmer, the sealer who sailed his 47-foot sloop *Hero* across the Drake, sheltered in Deception before casting southeastward across the unknown strait that was later called the Bransfield, where, perhaps, he was the first to see the coast of the great white continent.

Deception is the island to which many ships have fled from storms. It is also the island from which men and ships have fled, chased by nature's fire. Deception is the caldera of a snoozing volcano, which sneezed in 1967 and 1969. In 1921 the bay boiled, peeling paint from the hulls of whaling ships there.

Deception is a place of shelter and ruins. It's a perfect ring of rock, broken only by the narrow entrance of Neptune's Bellows. Within, the water is calm. Hot springs raise steam along the shoreline. Beside the faces on the tanks are rows of bleached and broken barrels left over from the old whaling station that once occupied the shore; on the other side of the tanks are buildings and the fuselage of an aircraft, abandoned by the British when the volcano got obstreperous. Farther around the island is an Argentinian base that is still occupied occasionally and was used as an emergency shelter for the crew of an Argentinean ship that ran aground in early 1989 near Anvers Island to the south. Not far from one of the hot springs are the stark, warped ruins of a Chilean base that was destroyed by the volcano's heat.

The U.S. Research Vessel *Polar Duke* once spent a windy afternoon working with a trawl to see what grew on the bottom of Deception's harbor. Bits of volcanic rock came up, but when the catch was dumped on the deck, it was startling. The deck squirmed with thin-legged starfish, thousands of them, slowly moving: a heap of unexpected life. The common name of the creatures reminded me of the fragile world of life and beauty in Antarctica. The floor of Deception is paved with brittle stars.

In the austral fall of 1988 a ship I was on left Deception Island and steamed north along the edge of the Bransfield Strait toward King George Island. It passed a great crescent bay of ice and rolled slowly up toward Livingston Island. One of the passengers, a young correspondent named Leslie Roberts, her yellow hair sticking up all over, came up on deck and looked beyond the pacing blue sentinels of icebergs to Livingston's skyline of iced mountains and windblown clouds. She was stunned. The scene matched old dreams. "When I was a kid," she said, "that's what I always thought heaven looked like."

YOU CAN'T GO FAR IN THIS PART OF ANTARCTICA WITHOUT COMING UPON SOME evidence of the presence of human beings. When I visited Spain's new base on the shores of Livingston in the fall of 1988, everyone was gone, but they had left crates of apples, oranges, and onions on the beach. On a post was a typical Antarctic sign on which people had tacked wooden arrows pointing to the cities of home, with the distances in kilometers: Madrid, 12,357; Cartagena, 11,475. The Spanish base had not been here long enough to acquire the other kind of human monument common to Antarctica, the wooden crosses of graves.

Sailing north from Livingston, the ship passes the forbidding ice cap of Nelson Island, where, in 1985, an aircraft carrying American tourists crashed while trying to land. The airstrip the plane never reached was on Nelson Island's neighbor, the far more hospitable King George Island.

Sooner or later anyone who goes to the peninsula's islands arrives at King George. The largest of the South Shetlands, it is notched with bays on the lee side of the prevailing Drake gales and offers ice-free shelves of rock where people can take a relatively easy foothold.

"We just think this place is very proper for beginners," said the commander of one of the newest bases.

The most striking thing any beginner notices about Maxwell Bay is the many faces of the nations that you meet on the shores. (The United States, which is decidedly not an Antarctic beginner, maintains a year-round base called Palmer Station on Anvers Island, roughly 65 miles south of the South Shetlands.)

TENIENTE JUBANY, ONE OF SEVERAL ARGENTINEAN BASES IN ANTARCTICA, IS THE first base to show up as a ship proceeds west from the Bransfield Strait into the shelter of Maxwell Bay. "La Gloria de Dios y La Honora del Hombre" begins the plaque on a monument high over the base, but the place does not look particularly blessed by God or honored by men. Its red buildings stand exposed like weathered outcrops, built on snowy gravel below a weathered volcanic plug called the Three Brothers. The only time I have ever been inside Jubany was in 1984, when a ship I was on delivered a psychologically damaged Argentinean to be sent home. But four years later I sat high on the rocks above the station, myself troubled by a long and contentious ship's journey. I sat looking at lichen growing in the lee of a rock the size of a football, and found that the stark landscape of stones and snowdrifts slowly engulfed me in the sanctuary of contemplation, the mingling of memory and present that has no regret, no hope, and no plan; nothing but the respect of the living for the mystery of the planet.

Jubany is a source of good water as well as reassurance; another time I watched and listened in on the radio as the Russian icebreaker *Mikhail Somov* worked a hose ashore to take on water, and its skipper discussed the operation with the Argentinians in their own common language: English. "Anything you need, anything you want," the Argentinean base leader said, "I will be on this frequency."

This was not an uncommon courtesy among nations in Antarctica. All the residents of this harsh landscape are on the same frequency, in a sense, coping with the weather and the loneliness together. As you leave Jubany and sail into Maxwell Bay, you pass bases built by a strange sequence of nations: South Korea, Uruguay, Russia, Chile, the People's Republic of China.

They are all here for the same reasons: to study an alien world, to position themselves in a possible international race for resources, and, for the people, to challenge themselves against a hard way of life.

Not long ago a South Korean magazine offered its readers a few exalted reasons for the construction of King Sejong base, which looks like a tidy derailment of a dozen orange railroad refrigerator cars. "The Antarctic," the article read, "mankind's last remaining treasure trove of rich natural resources, abounds in mineral deposits: oil estimated at 50 billion barrels, coal, copper, iron, uranium, gold, and silver. The krill supply in the Antarctic Ocean is more than enough to meet the protein demands of the entire population of the world."

It sounded familiar; it sounded just like what Yves-Joseph de Kerguélen-Trémarec wrote in 1772 after a glimpse of an ice-shrouded island: "No doubt wood, minerals, diamonds, rubies and precious stones and marble will be found. . . . If men of a different species are not discovered at least there will be people living in a state of nature, knowing nothing of the artifices of civilized society. In short South France will furnish marvelous physical and moral spectacles." Kerguélen-Trémarec was all wrong, too.

The recent international interest in mineral development in Antarctica, which culminated in 1988 with the signing of a convention that may allow companies to drill or mine there without settling the complex question of who actually owns Antarctica, has stimulated some of the base construction. Both the Uruguayan base, Artigas, a group of rudimentary huts slapped together on frostbound gravel between a blue wall of glacial ice and a black cliff, and the Great Wall, the Chinese base on the other side of the bay, were built in the past five years mainly to give their nations a voice in the Antarctic Treaty System.

The Chinese base was built in 1985 by 591 soldiers, sailors, and workmen. When I was there last year it housed a scientific and support crew of 16. Like many of the other bases, it was made up of several separate low buildings on short stilts. Some of the people there worked at the base for up to two years at a time. They were just like everyone else here—desperate for word from home.

"When we see the airplane," said one, "we rush over to it for letters."

The plane he was talking about was one of the regular flights into the hub of activity on King George Island, Chile's Teniente Rodolfo Marsh base. Marsh is a major airstrip and hotel offering small rooms with bunk beds and a view of the airstrip and the dusty wind. It reflects Chile's determination to maintain the appearance of owning a wedge of Antarctica, although the country's membership in the treaty system prevents it from officially asserting its claim. Marsh is connected by a mile of dirt road to Chile's other base here, Presidente Frei Montalva, which adjoins a hillside covered with long cream-colored buildings on stilts. The Chileans call this "Villa Las Estrellas." The most striking tactic in Chile's unstated campaign in Antarctica is colonization: Villa Las Estrellas is full of families. Twelve-year-old boys charge up and down the dirt roads on mountain bikes, and seven-year-olds play underneath the buildings. In 1984, when I first visited, there were four families and eight children in town; by 1988 the colonists, who each serve a two-year tour of duty, had grown in number to 12 families with 27 children.

A hundred yards from the Villa Las Estrellas, on the other side of a small stream, all these possessive machinations are watched with philosophical and socialistic detachment by the monastic men of Bellingshausen Base, which was built here in 1968 by the Soviet Union. Bellingshausen has always seemed a friendly place, although once I noticed its two underworked physicians looking speculatively at me in the hope I might be stricken. Outside Bellingshausen is another one of those Antarctic signs, only this one points home to Moscow, 15,200 kilometers away.

SOME LIVING THINGS FEEL AT HOME IN ANTARCTICA. IN FACT, ALTHOUGH MOST of the ice-covered continent is barren, the neighborhood of the South Shetlands teems with wildlife. Everywhere fur seals romp and bark; small armies of adélie, chinstrap, and gentoo penguins stand around in loose formation on small icebergs or crowd aromatic rookeries; thousands of cape pigeons, antarctic petrels, southern black-backed gulls, and other seabirds soar over the waves; and whales blow and breech. One day in calm water our ship paused in a southward run to shut down its engines and drift for a magical half hour while six humpbacked whales played around it as curiously as dolphins. That night we anchored far south, near the famous Lemaire Channel, a passage between vertical cliffs of islands and the mainland (a place so popular with tourists that it is also known as Kodak Gap) and were surrounded all night by the sounds of breathing: Seals patrolled the darkness.

Amid all this life, humans remain foreigners. At Chile's Captain Arturo Prat station on Greenwich Island, people deal with the oppressive isolation by building religious shrines and painting reassuring slogans on the rocks: "In the beginning God created Heaven and Earth." At Henryk Arctowski station, the Polish base in Admiralty Bay on King George Island, people use other tactics.

No one who has gone to the Antarctic Peninsula has ever regretted visiting Arctowski, though the recent growth in the tourist industry has made even this most hospitable place less welcoming. It is a place of unparalleled human warmth and incredibly powerful liquor. I have visited there three times, and it was circumstances alone that allowed me to remember leaving Arctowski each time. Some of my friends recall the toasts, but not the departures.

Arctowski has a miraculous structure—a greenhouse. These people who spend the winter in Antarctica's bleak whiteness go to breathe the scent of flowers and taste a ripe tomato.

"The greenhouse," said Piotr Presler, base leader in 1988, "is only for psychological condition."

Psychological conditions are close to the surface of all life in these islands. Once I sat at a long table in the paneled lounge and dining room of Arctowski, eating pickled gherkins and sausage and drinking Zywiec beer, listening to Presler, who spoke like a man tormented by his thoughts. "In many people is something . . . atavistic," he said. "Some men like to be in strong condition. It is part of why I come back here." At home he had a wife and a two-year-old child.

Presler looked sad. "After this year," he said, "I will go home and stay to the end of my life." Then he left an idea hanging over a hidden desolation: "If you know polar regions only from pictures," he said, "you say it is beautiful." And he changed the subject.

But José Augusto de Alencar Moreira, commander of the Brazilian base across from Arctowski, looked the subject in the face. His station, Commandante Ferraz, is built near abandoned British buildings at the bottom of a mountain, upon which stand three crosses. He knew that the challenges human beings face in this remote place are only partly caused by the cold.

"The most important thing is to understand each other," he said, "and have patience with the particularities of each other. It is one thing to say: 'I forgive you.' But then you put yourself in a position of power. It is much more important to comprehend than to forgive."

Beyond the islands of the edge of Antarctica, the ice of the peninsula itself rises like a spine and grows into the huge West Antarctic Ice Sheet. It looks like a continent. In one important way it is not. It is made almost entirely of ice, most of which rests on land far below sea level. Many scientists think the West Antarctic Ice Sheet is unstable. As the globe warms within its greenhouse of carbon dioxide, the ice may be sliding into the sea. So it is possible that within a few generations, if the climate warms and the seas rise, and the ice sheet slides away ever faster, that the whole of West Antarctica will turn into an archipelago, and a thousand new islands will emerge, fresh and clean and stony, from under Antarctica's most mysterious and magical face—the ice. And people will move to those islands, and try, as always, to comprehend not just the amazing landscape, but themselves. [1989]

North America

North to Alaska's ABC Islands: Admiralty, Baranof, Chichagof

KIM HEACOX

IT IS A TYPICAL SUMMER day in southeastern Alaska where life is anything but typical. A ninety-year-old man goes out in his skiff to catch a fish for his wife. A ten-year-old girl rides her bike down a dirt road past an Alaska brown bear. A bald eagle lands on the steeple of a Russian Orthodox church, as a raven calls from atop a Tlingit totem pole. A troller hauls in a ton of salmon. A logger fells a hundred-foot-tall hemlock. A kayaker paddles among a dozen sea otters, while not more than a mile away cruise ship passengers feast on fresh crab omelettes, play bingo, and watch for whales. And over in Juneau a pilot taxis a Cessna 206 floatplane into the wind and rises into the crisp, clean air of the Last Frontier.

"Juneau Tower, this is Cessna four-six-Alpha-Kilo," he radios in the terse lingua franca of pilots around the state.

"Four-six-Alpha-Kilo," comes the response, equally terse, "squawk zero-one-eight-eight." The Cessna climbs to 2,000 feet, banks smoothly, and heads toward Admiralty Island.

Admiralty, Baranof, and Chichagof—Alaska's ABC islands—form the northern vertebrae of the 1,100 islands in the backbone of the state's southeast panhandle, home of that famous marine waterway, the Inside Passage. They are an impressive triumvirate, these three islands. Roughly equal in size, each is strikingly wild. Each has colorful characters, abundant wildlife, a fruitful shore, and a mountainous interior. And each contains entire worlds of spruce and hemlock within the Tongass

National Forest, largest of the United States' 156 national forests. Chichagof and Baranof, on either side of the Peril Strait, together form the shape of an arrowhead; Admiralty lies just to the east across Chatham Strait.

No ordinary landscape, the ABCs are everything Alaskan, the kind of place Prince William Sound used to be—free of oil, rich in wilderness, and generous to those who make a living here. Like so much else in southeastern Alaska, these islands will charm you in the sunshine, chill you in the rain, and take your breath away at the slightest provocation.

Packed into the Cessna are myself, two talkative sportfishermen, a demure teenage Tlingit girl, a big bundle of camera gear, a bigger bundle of fishing gear, six dozen doughnuts, two pizzas, and the pilot. Our destination is Angoon, population 470, the only permanent settlement on Admiralty.

The island itself looms ahead, all 1,650 square miles of it. Add 400 square miles, and it would be roughly the size of Delaware. The native Tlingit Indians call Admiralty *Kootznawoo*, "fortress of the bears," for the estimated 1,700 Alaska brown bears that live here, an average of one per square mile.

In fact, a few are below us now, and the pilot banks the plane for a closer look. "There's three," he says. "Looks like a mama and two cubs." The bears move across a tidal flat, their fur cinnamon colored in the morning light. We swoop down and stare at them, and they don't even glance up.

Ten minutes later we land on the glassy waters of Mitchell Bay, next to Angoon. Smiling Tlingit faces greet us on the dock, smiling not for us, but for the doughnuts and pizzas—priority items in the Alaska bush. The teenage girl, so quiet in the plane, bursts to life when surrounded by her friends.

Come evening I find myself on the balcony of the Favorite Bay Inn just outside town, reading Hemingway and eating halibut as the sun swings around to the northwest and sets at 10:30 p.m. In the windless night the air is a prism, the water a mirror. A loons calls, then a bald eagle. I fall asleep to the music of Alaska.

"I TAUGHT SCHOOL IN SPOKANE, WASHINGTON, FOR 17 YEARS BUT GOT TIRED OF the city bureaucracy," Roberta Powers tells me at breakfast the next morning as she cooks berry jam on the stove. She owns and operates the inn. "So I came here six years ago and have been teaching kindergarten and running this place year-round." Roberta's husband, Dick, a former employee with the Forest Service in Juneau and Yakutat, owns Whalers'

Cove Lodge, a premier sportfishing lodge only minutes away by outboard skiff.

After a big bowl of cereal with handpicked blueberries, I head into Angoon, a ten-minute walk down a gravel road. This is a Tlingit town.

I stroll along, smiling and saying hello. Many reciprocate, but not all, for I am an outsider. I suspect the mistrust reaches back to October 1882 when a Tlingit shaman from Angoon was fatally injured while employed by a whaling company nearby Killisnoo Island. Because he was an important man, the Tlingits asked for the customary recompense: 200 blankets.

The whaling company, owned by whites, interpreted the Tlingit action as a threat and notified the territorial military forces in Sitka, on Baranof Island. A few days later the U.S. Navy arrived and bombarded Angoon, killing six children and destroying tribal houses, storehouses, and canoes, as well as treasured ceremonial hats, bowls, and blankets.

It was, said one Tlingit elder, "the day we paid for a crime that was not committed." That winter, with no shelter or canoes, the Tlingits of the settlement nearly starved. The pain lingered. One hundred years later, in 1982, the Tlingits asked the U.S. Navy for a formal apology. It wasn't the first time they had asked, and as before, they received none.

Carrying this history with me, I better understand the pride and pain of the people of Angoon. They still make their living from the sea, still maintain their clans of Grizzly Bear, Killer Whale, Frog, Beaver, and Wolf; they still smile, because life is good in southeastern Alaska, and they still expect an apology, because right is right and wrong is wrong.

Down by the waterfront I befriend half a dozen Tlingit children who are skipping stones. We talk, exchange names, skip more stones, then poke around tidal pools to see what lives there—urchins, limpets, sea cucumbers, sunburst starfish, and orange anemones. One boy pries a limpet off a rock with a jackknife, scrapes out the soft body, and eats it. "When the tide is out," the Tlingit say, "our table is set."

LATER THAT DAY THE STATE FERRY *LECONTE* ARRIVES, DISEMBARKS SOME PASsengers, loads others, including me, and begins the three-hour trip up Chatham Strait to Tenakee Springs, on Chichagof Island. For the next two weeks this is how I will see the ABCs, in A-C-B order, traveling by ferry, floatplane, sailboat, seiner, and fish packer, following no set schedule, just moving when it feels right or the opportunity arrives. "Where ya headed?" fishermen ask, and when I tell them, they say, "Well, we're goin' there in the morning. Yer welcome to ride along if ya like."

From the decks of the *Leconte*—one of the smallest ferries in the state fleet—I watch Admiralty slip by to starboard and Baranof to port, their shaggy, green slopes tumbling into the sea. Passengers relax in deck chairs on the solarium, have a hot meal in the cafeteria, strike up conversation with new friends.

Most visitors to southeastern Alaska make their marvelous journey along the Inside Passage by luxury cruise ship or large state ferry. Yet each year more jump on small ferries, boats, and planes to see a different slice of Alaska: Angoon, Tenakee Springs, Hoonah, Pelican, Elfin Cove. Of the dozen or so communities on the ABC islands, only Sitka is a regular cruise port of call.

The hundred or so people of Tenakee Springs would probably faint at the sight of a cruise ship dropping anchor at their town. There are no roads, no cars, no newspapers, no banks here. A single path, flanked by homes built on stilts over the sea and against the hillside, parallels the waterfront through the tiny town. A sign in Snyder Mercantile reads, "If we don't have it, you don't need it." The post office is slightly larger than a closet; the postmaster only slightly smaller than Paul Bunyan. Time takes its time here. People stroll more than walk. They greet each other by first name and ask about important matters like last night's card game. In Rosie's Blue Moon Cafe, reputed home of the best french fries in southeastern Alaska, two men discuss the merits of chocolate versus vanilla ice cream. And at the Tenakee Inn fishermen drink beer, roll their own cigarettes, and tell stories into the night.

"I came here ten years ago for a part-time job," a bearded man named Bushman Jack tells me. "I didn't get the job, but I liked the place and stayed. There's not much work, Tenakee being a sorta retirement town. I do odd jobs, whatever it takes to get by—fix a boat here, paint a house there. It's slow, but I like it that way. People who want a faster life can live in Anchorage, or Seattle, or New York. I live in a cabin on the outskirts of town, cut my own wood, haul my own water, and sew my own clothes. I walk to work and haven't locked the door or had nothin' stolen in ten years. That's the kinda town this is . . . friendly. Can I buy you a beer?"

THE FOCAL POINT OF TENAKEE SPRINGS IS THE NATURAL, SULFUROUS HOT springs itself, at the head of the dock, with men's and women's bath hours alternating around the clock. Were it not for that, the town would not be here. Tlingits showed Klondike miners the springs back in the 1890s, and people have been coming here ever since. The town popped up but never grew much. Now, any time of the day or night you might see folks shuffling down the path with towels over their shoulders on their way to the best therapy in town. They might be locals taking a dip for the 500th time, or visitors like me, going for the first.

I slip into the hot water and relax, and by the time I get out, still radiating with heat as I walk back down the footpath, Tenakee Springs is on my list of places to visit as often as possible. The experience is so nice that later, after reading Oscar Wilde until 2:00 A.M., I leave my room at the Tenakee Inn and return to the springs for another soak. As Wilde said, "I can resist everything except temptation."

I am not alone. Several Tenakeans, including Bushman Jack, are already in the hot water talking about the only thing hotter: the "Tenakee Road issue." The roots of the controversy reach back to 1980, when President Jimmy Carter signed the Alaska National Interest Lands Conservation Act, creating more than two dozen national parks, preserves, monuments, and wildlife refuges in Alaska. Among them was Admiralty Island National Monument, most of which was awarded "wilderness" status and made off-limits to logging. In return, however, Alaska's congressional delegation demanded and received a special clause in the legislation whereby the Tongass National Forest, in a unique arrangement, would receive a guaranteed $40 million per year to survey trees, write documents, build roads, and make available for cutting an average of 450 million board feet of timber per year for the next ten years. Critics say the Tongass is a disaster ecologically and economically. The U.S. Government is losing money—an average of 73 cents on the dollar—to finance the clear-cutting of the temperate rain forest of Alaska.

"The Alaska Chainsaw Massacre," grumbles a man in the hot springs next to Bushman Jack. Northeast Chichagof Island is not nearly as pristine as it once was, he says. Clear-cuts fill the drainages, brown bears are shot on sight, roads radiate from Port Frederick and the town of Hoonah. One of those roads, if extended another mile, would connect into Tenakee Springs and bring tremendous change.

THE NEXT DAY I PACK MY BAGS AND HOP A SAILBOAT TO HOONAH. UNLIKE Tenakee, Hoonah is a noisy, hard-working, greasy sleeves town—the largest Tlingit settlement in southeastern Alaska. Slogging down main street in the pouring rain, I meet loggers, fishermen, children and parents, Tlingits, whites, Asians, ten cats, a parrot, and at least 40 dogs, all in less than an hour. Six loggers at the opposite end of a restaurant bar order hamburgers for lunch, so I do the same, thinking here's a chance to spark a discussion about the Tongass National Forest. But given the loggers' size and their comments about "crazy environ-maniacs"—and my interest in staying alive—I say nothing. Besides, it's impolite to speak with your mouth full. Slipping outside, I go in search of a kinder, gentler Hoonah.

Many years ago at a lodge in Glacier Bay, across Icy Strait, my path crossed that of George and Jessie Dalton, an elderly Hoonah couple who spoke Tlingit. For one memorable evening I listened to their translated stories.

One, I remember, told of a young girl who went berry picking alone. "Be careful of bears," her mother told her. "Bears are dangerous." The young girl went out, filled her basket with berries, then sat down and fell asleep. When she awoke, her basket was knocked over and licked clean, and she was surrounded by bear tracks, the biggest she had ever seen.

"Is that a true story?" I asked George.

"Of course," he said.

I inquire around town about George and Jessie, hoping to see them again, and finally find a Tlingit man who tells me, "Oh yes, they were in poor health for awhile but are feeling better, I think. They'll be celebrating their 70th wedding anniversary soon. They're out of town right now. Probably picking berries."

THE *LECONTE* TRAVELS DOWN LISIANSKI INLET AND VISITS THE FISHING TOWN of Pelican only twice a month. "Special events here," reads a notice in the laundromat, "include the arrival and departure of the state ferry, the tide change, sunny days, and a woman in a dress."

At the Pelican Wet Goods and Steambath I ask owner Harry Owens how often he gets out of Pelican. "As seldom as possible," he says. "There's no place I want to go. I don't like Juneau, don't like Sitka. Can't stand any place down south. I've been here 18 years. Pelican suits me just fine, and then some."

Founded in 1938 and named for an early fish-packing vessel, Pelican bustles as trollers and packers deliver salmon, halibut, crab, herring, and black cod from Icy Strait and the Fairweather Ground in the Gulf of Alaska to the seafood processing and cold storage plants. "Closest to the Fish," reads the sign at the head of the dock, for this is a staging area for the largest commercial fishing fleet in southeastern Alaska.

The season happens to be a banner year for pink salmon. "The best since 1940," a fisherman tells me on the waterfront as he cleans his fingernails with a knife. "I've never seen 'em so thick." Later that night in Rosie's Bar and Grill, he leans over and says, "See that skipper over there wearing the beret and buying all the beer? I'll bet he's cleared $10,000 in the last two days."

"Yeah," another fisherman says, "and he'll spend it that fast, too."

What charms me most about Pelican is the boardwalk. Raised on pilings over the tide, it ties the town together. People meet here, lean against the rail, and tell their stories. Or, like me, they stop to watch spokes of sunlight play on the mountains, forests, and water.

"If you like Pelican," a woman tells me, "you gotta go to Elfin Cove.

It's the most enchanting little town I've ever seen." That evening a fish packer pulls away from Pelican with several tons of ice and one writer on board, bound for Elfin Cove.

Enchanting is an understatement. On the north end of Chichagof, across Icy Strait from the shining Fairweather Range, this little town, population 50 (give or take 30, depending on the season), is sheltered by a peninsula and several islands. There's a narrow, winding boardwalk from which three separate docks branch into the calm water. Add to that a cluster of rustic homes, a store, an inn, a post office, three sportfishing lodges, the smell of the sea, and the laughter of children, and you have it: Elfin Cove.

The fog rolls in and out, the boats come and go, and I can hear the fishermen singing in the mist as they off-load their salmon. Youngsters climb in skiffs and go out fishing themselves, or just drop a hook and line off the dock. For hours I sit there and watch the masts move across the amber light of dusk and dawn.

"Some days I pinch myself to make sure all this is real," says Dennis Hay, owner of Elfin Cove Sportfishing Lodge. He and I stand together on the lodge balcony, taking photographs. And as the sun sets behind the Fairweather Range and a bald eagle flies across the crimson sky, Dennis turns to me and says, "Do you honestly think a photograph can capture this?"

TWO DAYS LATER I'M OFF FOR MY FINAL PORT OF CALL: SITKA, ON THE WESTERN coast of Baranof Island. Called the "Paris of the Pacific" before San Francisco stole the title in the mid-1800s, Sitka is still the siren she used to be. Nothing like a pretty town on a wild coast to steal your heart away. Sitka feels cosmopolitan compared to Elfin Cove and Tenakee Springs, and after two weeks in those quiet corners I am ready for paved streets, bookstores, art shops, and music festivals; ready to meet people off a cruise ship, to buy a gift for my wife, to sit and contemplate a Tlingit totem pole, and to walk reverently through St. Michael's Russian Orthodox cathedral.

History burns brightly in Sitka. Here in 1802 Tlingit warriors battled Russians and the tides of change, and lost. Here the indomitable Aleksandr Baranov built the capital of the Russian-American Company. And here in October 1867, with the furs gone and the sea otter on the brink of extinction, the Stars and Stripes was raised atop Castle Hill to commemorate the $7.2 million purchase of Alaska negotiated by U.S. Secretary of State William Seward. "Seward's Folly!" cried an angry American public. "What good is Alaska?"

Climbing to the top of Castle Hill, I stand among the cannon, the alder, the old stone walls. In one direction rise the rooftops of Sitka, backdropped by steep mountains. In the other direction lie islets sprinkled before the open Pacific. Tenders run to and from a cruise ship anchored in the harbor as a light rain falls. It must be the sixth or seventh time I have climbed Castle Hill, and as I have every time before, I take a moment to give thanks that I live in Alaska. Good for you, William Seward!

ADMIRALTY, BARANOF, AND CHICHAGOF ISLANDS FADE OFF THE STERN OF THE ferry heading to Juneau. It seems more fitting to say goodbye this way, drifting away on a boat rather than blasting away on a jet. I sit topside, where the air tousles my hair. A floatplane flies overhead, a whale surfaces nearby, and a seiner hauls in salmon. It's a typical day in southeastern Alaska.

"Sure is incredible country," says a white-haired man next to me, almost as if talking to himself. I smile and he adds, "You know, if I were young like you and coming into this country for the first time, I doubt I'd ever leave." [1990]

Vancouver: Seeking a Natural Balance

BILL BARICH

THE WEDDING IN SEATTLE was on a Sunday, but the party had started on Saturday afternoon, and by eight o'clock Monday morning, when I boarded the ferry for Vancouver Island, I had the sort of headache that causes sailors to invent mythical stories about the vindictiveness of the sea. Gray, drizzly clouds had clamped a lid on Puget Sound, and the water everywhere was laced with whitecaps. Collapsing into my seat, I heard the soppy splash of waves against our hull. The ferry, freed from its pier, began rocking gently and then not so gently, tilting at such a perilous angle that the marine landscape bobbed and down like a boxer ducking punches. For a moment or two I thought about grabbing my gear and leaping back toward land, but the engines finally kicked in, and the ferry seemed to right itself.

We moved smoothly out of the harbor, slowly gathering speed until we were traveling through an inlet that offered us some classic views of the Pacific Northwest. Gulls sailed over deep green firs and spruces veiled in an atmospheric blend of mist and fog. Past Whidbey Island we went, and past Port Townsend, and into the open water of the Strait of Juan de Fuca. On our port side I could see the massive, thickly timbered peaks of the Olympic Mountains, bluish black against the slate-colored ocean.

After a while I tested my sea legs by taking a spin on the passenger deck. Through the haze the largest island in the North American Pacific was visible in the distance. Vancouver Island is about 280 miles long, and its topography is so rugged that it has never been densely populated, except around Victoria, the provincial capital. A mountain range divides the island into two distinct regions—an eastern plain that is sheltered and has a relatively mild climate, and a more weatherbeaten, rain-soaked west coast, where a handful of communities sustain themselves by practicing the traditional island industries of logging and fishing.

The west coast was my destination. I was hoping to do some hiking and exploring in and around Pacific Rim National Park near the towns of Ucluelet and Tofino. As an added attraction, there would be a chance to observe the annual migration of gray whales from their calving grounds in Baja California to their summer home in the arctic seas. And though it was only April, I'd brought along a fly rod, knowing that I wouldn't be able to resist temptation if I happened on a likely looking trout stream.

When we reached Victoria, I hired a bearded cabbie in a turban to give me a tour of downtown before dropping me at a car rental office. We rode through an extremely clean city that was as prim and self-consciously pretty as an English seaside town. Victoria became a crown colony in 1849, and its British heritage is still eminently apparent, but it feels bland and neutral, lacking in character. Although modern condos are going up near the water, most of the city's architecture harks back to the 19th century, especially the stone Parliament buildings and The Empress, a fantastic dowager of a hotel done in such brash colonial style that it employs some Indian waiters apparently schooled by Rudyard Kipling to serve drinks in its Bengal Lounge.

Tea parlors. Pubs. Shops selling Harris tweeds. There was even a Marks & Spencer department store that featured the same goods you might find at a London branch: pork pies, tinned peas, and the unfashionable cardigan sweaters that retired majors like to wear while reading accounts of military battles by the fireside. The British influence, faux and otherwise, was so absolute that I wondered if Victoria had an identity of its own, one that was purely Canadian.

So, while picking up my car, I put the question to the clerk, whose name might have been Desmond or Derek or Basil. He'd been reading about the Stanley Cup play-offs in a newspaper, hockey being the arena in which Canada's ordinarily repressed violence gets acted out.

"Oh, I wouldn't know about that now, would I, sir?" he said, as though he'd been asked to expound on Hegelian logic. Tossing me the keys to a dusty Grand Am, he showed me the route to the west coast on the map. It was nearly 200 miles away, so I chose to spend my first night in the logging town of Port Alberni.

I COULD SMELL PORT ALBERNI BEFORE I COULD SEE IT. IT HAS THREE MILLS that turn out lumber, specialty cedar, plywood, and pulp, and they scent the air with a pungent, chemical odor. On Alberni Inlet, where the landscape resembled Norwegian fjords, some lumber freighters and fishing boats were docked. The inlet and nearby Barkley Sound are famous for producing Chinook salmon, yielding about 20 percent of British Columbia's total sport catch, and every summer some anglers hook a few really huge ones—tyees that are the whales of the salmonid family, tipping the scales at more than 30 pounds.

Tired from my day on the road, I settled into the dining room at my motel and had a surprisingly good meal of thinly sliced smoked salmon and a Caesar salad. Three local men were nursing pints of beer in a little alcove bar decorated with Reader's Digest Condensed Books, and though they spoke in the hushed, almost sedated tones that are common on the island, I was able to listen in.

They were grumbling about the town of Tofino—about its pricey restaurants and its expensive waterfront homes, about the tourists, the developers, and the hippies. They made it sound like an unlikely combination of Puerto Vallarta and Gomorrah, and I began having wildly improbable fantasies about hammocks, daiquiris, and the beautiful Tofino maidens who'd be strolling the sandy beaches at twilight in their Canadian-style sarongs.

The next day I took off for the west coast on Highway 4, which was completed in 1971. Before then Ucluelet and Tofino were truly isolated, and to get to them by land you had to negotiate rutted dirt or gravel logging roads, while simultaneously dodging the many speeding trucks that appeared to be bent on your destruction.

As I skirted the edge of Sproat Lake and climbed into the Mackenzie Range, the sun peeked out between flocky clouds, and for the first time I felt that I was seeing the Vancouver Island of postcards. The lake sparkled, birds flitted through the forest, and brilliantly crystalline creeks flowed over pebbled creek beds. Toward Sutton Pass summit the clouds closed over the sun, and a light snow began falling, clinging to the branches of trees and dusting the fronds of delicate adder's tongue ferns.

The trip would have been perfect, in fact, if it hadn't been for all the ugly, stump-ridden clear-cuts marring the woods. The timber industry has always been the boss on Vancouver Island, but in the past few years there's been a serious backlash against its most destructive practices, such as clear-cutting.

This year, for instance, a German TV documentary, *Decaying Paradise*, embarrassed the government of British Columbia by comparing the loss of its forests to the more publicized destruction in Brazil. But logging is so deeply entwined in the island's economy and its history that unraveling it presents a problem—one that pits blue-collar families like those in Port Alberni against people whose incomes aren't dependent on timber harvesting.

The highway came to a dead end near the ocean, and I turned onto a road that led south to Ucluelet. Surrounded by logged-over hills, Ucluelet was originally a Nuu-chah-nulth Indian fishing village. Fishing is still central to the town's welfare, but its most prominent asset is the Canadian Princess Resort, a fairly new hotel complex built around a docked ocean liner. For about $70 a night you can stay in the captain's suite on the *Canadian Princess*, but it was already reserved when I checked in, so I had to put aside my thoughts of sipping brandy from a flat-bottomed ship's decanter and playing cribbage with the purser. Instead, I bunked at the hotel proper where, after a hearty clam chowder, I spent the evening reading about the whales I was going to pursue in the morning.

DAWN IN UCLUELET BROKE CALM AND ROSY. AT NINE O'CLOCK I WAS STANDING at the hotel's dock among hordes of French-speaking schoolchildren and several groups of elderly folks, ready to board the *Nootka Princess* for a whale-watching trip. The *Nootka* was a sharp little vessel rigged out for salmon fishing, but the whales were bringing in some real money on the coast, and it wouldn't have shocked me to see a bathtub equipped with an outboard motor pressed into service. As we pulled away from the resort, our skipper, Captain Mike, announced over the PA system that we were bound for Barkley Sound, where hundreds of ships had gone down over the years, buried beneath the swells.

"'Graveyard of the Pacific,'" Captain Mike said cheerfully, and several passengers reached for their Dramamine.

While we chugged through a harbor channel lined with canneries and a few turn-of-the-century houses on stilts, I chatted with Carol, our first mate, who'd hitchhiked to B.C. from Niagara Falls and had stayed on in Ucluelet, earning a living by working on fishing boats. The love of the sea was strong in her, and she'd just bought her own dinghy for 500 bucks to use for summer excursions to some of the hundred or so Broken Group Islands in Barkley Sound. Her last boyfriend had been a fisherman, Carol said, and she'd spent ten months with him fishing from Vancouver to Port Hardy.

"We caught a lot of dogfish," she told me. "Have you ever smelled one of those?" She made a face suggesting that I wouldn't enjoy the aroma. "Ten months on a 24-foot boat bringing in the dogfish—it tends to ruin a romance."

When we were out of the channel we ran hard, bouncing over the blue-green waters of the sound, and then came to rest at a spot where some gray whales had been sighted earlier in the week. It's amazing that gray whales are still around, since they've been hunted to the brink of extinction more than once. It wasn't until 1946, in fact, as they were about to vanish from the earth, that they were classified as an endangered species. Grays are prolific breeders, though, and they soon grew in number and lost their endangered status. Nobody is certain how many of them migrate along the coast every spring, although some experts put the estimate as high as 20,000. The whales pause to feed on the way, dining on bottom-dwelling crustaceans at a tremendous rate. The point of the gorging is to add some new blubber, from six to twelve inches' worth, to get them through the winter months.

In a few minutes we saw our first whale sounding about 30 yards away, off the starboard rail. It looked more black than gray, and its hide was covered with barnacles. A mature gray whale like this one weighs between 20 and 40 tons. Even so, it moved with unexpected grace.

It disappeared almost as quickly as it had appeared. Then there was a period of hushed waiting before we heard a loud noise that reminded me of somebody trumpeting on a conch shell (though magnified many times), and a spout of vapor shot into the sky. Again the whale surfaced, this time accompanied by its mate. It was lovely to see the pair of them cavorting, and marvelous to think how comfortable they were in the vastness of the ocean.

All morning we kept seeing whales, some up close and some at a distance. They would surface, then dive, then surface again, caught up in a process that had some poetry in it. You couldn't watch them without feeling a profound admiration for the species, along with an amazement that they had managed to survive so much depredation. When we returned to the resort at noon, sunburned and with eyes stung by the ocean glare, I felt as if I'd been witnessing fragments of a primal dance that had previously been hidden from me—a dance whose very existence was a kind of gift.

TOFINO TURNED OUT NOT TO BE GOMORRAH, OR EVEN PUERTO VALLARTA. Instead, it was a picturesque village located where the Esowista Peninsula stretches into Clayoquot Sound. Against a backdrop of snowcapped mountains, many islands were floating in the blue, all of them dominated by Meares Island and Lone Cone peak. I found no developers and no beautiful maidens in sarongs, but I did bump into some hippies gathered outside a natural food store, where the staff had the blissfully arrogant attitude of the organically correct. The village also had a surf shop, a T-shirt shop, and a place called the LA Grocery, where you could buy Zig-Zag cigarette papers and neon-colored fanny packs. From the evidence at hand, it seemed that Tofino was enduring the initial stirrings of trendiness.

But that wasn't the primary reason why its neighbors in Port Alberni and Ucluelet frowned on it. In the past decade or so Tofino has become a symbol of Vancouver Island's anti-logging movement. In 1984 a coalition of environmentalists, Clayoquot and Ahousat Indians, and some local citizens blocked loggers from a multinational corporation from clear-cutting on Meares Island. The town's resistance to most timber harvesting has continued in the courts since then.

Because multinationals don't have an impressive record when it comes to resource management, Tofino residents worry that they'll foul the town's watershed, damage its salmon fishery, and destroy some of the last temperate rainforest belts on the planet. Not incidentally, the spectacle of clear-cut mountains might also drive away the tourists who travel to the Esowista Peninsula because it's still pristine.

Tourism has been an increasingly important source of income for Tofino since Pacific Rim National Park opened. The huge park, which includes the Broken Group Islands and the West Coast Trail, has everything from a golf course to an abandoned gold mine. Except in the summer high season, it's remarkably uncrowded, still wild enough to provide habitat for game birds, deer, black bears, cougars, and mountain lions.

I visited the park a few times and did most of my hiking around Long Beach, where the white sand is hard-packed and you can walk for miles along the ocean. The beachcombing was excellent, with driftwood, shells, and Japanese glass fishing floats washing up. Kite-fliers were out, and so were kids tossing Frisbees to their dogs. There are trails in the park that wind through rain forests and swampy bogs, and when I strolled them alone, with no one else in sight, I got a little unnerved at a rustling in the underbrush. It seemed preposterous that I'd encounter a bear at the beach, and yet in town I'd heard a tale of a surfer who'd bumped into one on his way to his van and had to scurry up a cedar tree.

Most evenings in Tofino I stopped for a glass or two of Okanagan Spring Pale Ale at the Blue Heron Pub. It's a friendly spot, where fishermen off the boats come in for a drink, and you can usually order some extremely fresh Dungeness crab that's cracked and served cold with mayonnaise. One night I had a talk with the bartender, Greg, who'd recently graduated from the University of Manitoba and lived on a nearby island, Wickaninnish, that his family owned. Greg was an avid angler, and he claimed that he could push away from Wickaninnish in a rowboat, drop a jig into the water, and almost instantly hook a salmon, letting it pull him around for 30 minutes or so before he landed it.

SHARON PALM LIVES ON HER OWN PERSONAL ISLAND, TOO, ALTHOUGH STRAWberry is much smaller than Wickaninnish. She operates a water taxi—a skiff with a 55-horsepower Evinrude motor—and I had hired her to take me to Meares Island, where I'd heard there's a terrific hiking trail. Sharon came to Tofino during the early 1970s, when the town was a counterculture haven. Now she has a husband and five children and makes her home in an old wooden ferry that's been reconditioned and is up on blocks. Her taxi business has given her an intimate acquaintance with Clayoquot Sound, and she offered me a running commentary, pointing out Morpheus Island, where Tofino's dead used to be buried (her family helps clear the blackberry brambles from the cemetery once a year) and Stockham Island, where the sick were tended. Stubbs Island, site of the first Clayoquot trading post, was farther away, just a speck on the horizon.

We went through some shallow water that barely covered a mud flat. At low tide, Sharon said, there were sometimes thousands of sandpipers feeding on the flat. As we got closer to Meares, she nodded toward a group of houses on the shore. This was Opitsaht, a Clayoquot settlement, where about 40 Indian families live. The island is a Clayoquot reserve, and there are shell middens scattered everywhere.

In the old days, Sharon told me, the tribe would have had two settlements, and the Indians would shuttle between them depending on the weather, fishing for salmon during the milder months and then moving to a quiet inlet for the winter, when hundred-mile-an-hour gales whip through the sound. Because Meares has such significance for the tribe, its members have been instrumental in the fight against logging, in spite of the much-needed profits it would bring them.

Sharon cut the throttle, and we drifted toward the island. It's a substantial piece of property, bigger than Bermuda. The trail was something of a disappointment, though; too muddy from the recent rains to permit any decent hiking. Sharon was afraid that I'd slip and break a leg, but she let me get out for a while to wander around. I took a few steps on the trail, feeling ooze creep over my boots, and stood in a grotto formed by the towering trees, the sort of damply fecund forest that gives rise to stories about pipe-smoking gnomes and enigmatic caterpillars. There was a constant trickle of water, *drip-drip-drip*, and hazy light filtered through leaves and branches to flood the forest floor. It was a magical moment, very peaceful and restorative, and I left the trail reluctantly.

MY FINAL DAYS ON VANCOUVER ISLAND WERE SQUANDERED IN A SEARCH FOR some trout fishing. Again I drove over the Mackenzie Range, but instead of going to Port Alberni I pulled into Parksville, a resorty town on the east coast. Only in Canada, I suspect, will you find such time-warp towns, where an elaborate miniature golf course constitutes the main drawing card. There were advertisements posted everywhere for the first annual Brant Festival, which appeared to be a celebration of the countless brant geese passing through the Strait of Georgia on the way to their northern breeding grounds. Parksville was an utter symphony of flapping and honking. At my hotel I asked a young desk clerk, a transplant from Glasgow, to tell me more about the events that were scheduled.

"I'm not sure," she said. "We've never had the festival before."

"You figure the geese know you're celebrating their arrival?"

"It's better than shooting them, I suppose."

At a tackle shop I bought some flies in exchange for some fishing information. "Try the Englishman River," the proprietor said. So I did. The river wasn't far from Parksville, but it was low and clear and not producing any trout.

The other suggestion was to head toward Lake Cowichan, about an hour or so north of Victoria. I turned off the highway at the road leading to Cowichan and passed through yet another patch of logged-over land.

The spectacle of trashed forests was really beginning to depress me, and Cowichan, too, was a lumber town and so was Honeymoon Bay, which, in my fantasies, I had imagined as a cluster of red-and-white cottages beneath some sheltering pines. I was about to quit and just drive to Victoria when I saw a sign for the Sahtlam Lodge and followed a dirt road. Then a miracle occurred.

The lodge was magnificent. It sat right at the edge of the Cowichan River, a stream that's noted for salmon and steelhead trout. A newspaper tycoon had built the lodge as a private enclave many years ago, but now it belonged to a young couple, Val and David Hignell, who had refurbished it with antiques and Japanese prints. There were three rooms upstairs, each outfitted in a style that wasn't the least bit precious. I booked one and set about an evening's fishing.

David, a vaguely academic fellow with a neat moustache, was not an angler himself, but he recommended a particular riffle on the other side of the river.

"How do I get to it?" I asked, looking for a footbridge.

"Use the cart."

The cart was a wooden crate suspended on a system of ropes and pulleys. I climbed into it, then released a latch and went zooming out over the Cowichan.

About halfway across, the cart stopped, and I had to yank on the ropes to drag it to the opposite bank. The entire procedure caused an immense amount of adrenaline to race through me, especially during the time I was suspended dizzily above the raging water of the river. No doubt the whole tricky business affected my ability to fish, because I hooked and threw back only one little trout. Ever a gentleman, David afforded me an excuse—too early in the season—and, without the slightest shame, I accepted it.

Val does the cooking at the lodge, and she does it with panache. The dining room tables are on a porch overlooking the stream, and when David brought me the first course of curried soup and homemade bread, I popped the cork on a bottle of Pinot and made a silent toast to the river.

The main course was rabbit sautéed with garlic and vegetables, and it was sublime and also plentiful—so plentiful I had to skip the salad and dessert.

Around nine o'clock I retired to my room, opened the windows, and watched the last light fade outside. I couldn't imagine a better way to end my trip. The pleasures of the lodge gave me an optimistic feeling. There was, I decided, no reason why Vancouver Island couldn't address its environmental problems before it was too late. The gray whales had survived, after all, and it was still possible to hope that the island's forest majesty would be accorded the same fate. [1991]

Lake of the Woods

JEFF SPURRIER

"THIS IS REALLY NORTHERN territory," says Bill Oemichen, easing back on the throttle of his Mercury 50 to let the boat coast silently into the weeds at the shallow end of Deep-water Bay. "Try a Red Devil."

In between slaps at deerflies, I rummage around in my tackle box, searching for the red-and-silver lure with the triple hook that Northern pike die for. Literally. I toss my line out to the edge of the marsh where lake grass mingles with wild rice, and slowly crank in. The sun has just dropped behind the hills of Falcon Island; twilight is folding around us in a skywide backdrop of magenta, pink, orange, burnt umber, and blue.

It's nearly 10 P.M. here in the Canadian waters of Lake of the Woods, and again I'm reminded how life is played out with different rules in Northern territory. We have charts, but distance on the lake is measured in islands, and time itself is suspended as the sun crawls across the heavens, stretching out each afternoon to childhood dimensions.

The air is still except for the threnody of bullfrogs in the weeds, while off toward shore a lonely great blue heron stands patiently, watching our mute progress through the water. The cry of a loon floats across the lake, as Bill points at a dark shape at the edge of the water some 200 yards away.

"A young moose," he whispers, "a bull. He's going after the weeds."

I'm about to reach for my camera, when suddenly the end of my rod snaps down violently, and the line goes whipping out. Fifty feet away a huge fish breaks the water, and I jerk back on the rod, adjusting the drag on my reel so he won't snap the line. After five minutes of fighting, the fish finally is close enough to be netted, and Bill hauls him on board—a six-pound Northern. After two frustrating days of casting, experimenting with everything from fresh leeches to Lazy Ikes and Deep Hogs, I feel that I've finally arrived at Lake of the Woods.

TELL A TRAVEL AGENT THAT YOU'RE GOING TO LAKE OF THE WOODS, AND YOU'LL probably receive a blank stare in return. For years this massive body of water, bordered by northern Minnesota and the Canadian provinces of Manitoba and Ontario, has been a backwoods secret shared only by geographers and midwestern fishermen. Lake of the Woods' 2,000 square miles of water make it the largest freshwater lake in the United States after the Great Lakes.

It's easy to forget all that water because of the islands—more than 14,000 of them. (In fact, when the first European explorer, Jacques de Noyon, stumbled out of the forest in 1688, he named his discovery Lake of the Islands.) In this northern landscape where 100-degree-plus summers and minus-40-degree winters are common, man moves in cautious circles, taking his lead from the wildlife that inhabits the islands: deer, black bear, timber wolf, moose, and beaver.

It was the beaver that brought to the area the first major incursion of European explorers, the voyageurs who canoed among the islands in the 1700s, trapping, trading, and searching in vain for the fabled northwest passage to Asia. A French expedition in 1732 established Fort Saint Charles on an island in the lake, but it was abandoned 17 years later after the death of the group's leader. The fort has since been reconstructed on Magnussen Island.

For the next hundred years the majority of the new settlers around the lake were Ojibwa Indians, pushing their way west into what had been Cree, Monsonis, Assiniboin, and Sioux territory. Until the late nineteenth century, the lake remained hidden behind miles of wilderness. Indeed, the first permanent resident of Lake of the Woods County in Minnesota didn't arrive until 1885, a fisherman named Wilhelm Zippel.

Today most of the lake lies in Canada, but a third of it is American, the result of an eighteenth-century mapping error that also created the Northwest Angle, a trapezoidal piece of Minnesota north of the 49th parallel, bordered on three sides by Canada. While Lake of the Woods deserves a Trivial Pursuit question (the Northwest Angle is the most northerly point in the contiguous United States), what makes it noteworthy for sportsmen is the fishing.

At least that's what I was told when I was invited to join Bill Oemichen (a former soil conservationist for the U.S. Department of Agriculture and now a farmer in north-central Minnesota) at a fishing camp on Oak Island, one of two large American islands in the lake, located a mile from Ontario's western border. The Minnesota Department of Natural Resources backed him up, citing the lake as the easiest place to catch your limit. These waters are home to Northern pike, muskellunge, sturgeon, and sauger. But the lake is most famous for a savory, if unappetizingly named, member of the perch family known as walleye.

Having grown up a saltwater fisherman in southern California, I wasn't sure if the trip would be worth it. Fishing for me has always meant heavy tackle, palm-size live bait, and thrashing fish that have to be gaffed. My initial feeling was that a four-pound line and a pencil-thin rod was, well, a little wimpy. Then I looked up walleye in my *McClane's New Standard Fishing Encyclopedia* and read that for its size, fighting ability, and flavor, "It is questionable whether any species has greater angling value."

ALTHOUGH A FEW HARDY INDIVIDUALS BRAVE OAK ISLAND'S WINTER AND COME to ice fish, most regulars wait till the end of May, and the spawning of the walleye. During the few weeks before school lets out and the families arrive, Oak Island seems to re-create *Field and Stream* images of the 1950s: sputtering boats reeking of gasoline, men with grizzled beards, huge tackle boxes stuffed with colorful, esoteric lures and nasty barbed hooks. The atmosphere crackles with outrageous lies, salty curses, and an endless inhalation of beer.

As I sit in the shuttle boat from Young's Bay to Oak Island, under a deep blue cloudless sky, the lake seems like it belongs more off the coast of Queensland, Australia, than northern Minnesota. It's hot, hot, hot. The captain of the launch tells me the ice has only been off the lake for a few weeks, and he can't explain the 100-degree temperature.

Still, he says, "the fishing's been great." As the ice clears, he explains, the walleye move into the shallow waters to spawn, making it possible to catch your daily limit of ten in just a few hours—not only small males but also hefty females, ranging from 3 to 20 pounds.

As the boat cruises across the still water, disorientation sets in. It's nearly impossible to tell what's mainland, what's island. The amoeba-shaped chunks of land sticking up out of the water range in size from tiny reefs to sloping shoulders of worn-down mountains thickly mantled with broad expanses of spruce, white pine, trembling aspen, and birch. They turn the lake into a confusing maze that makes navigation difficult.

"These are what the glacier left," says Bill, gesturing to the islands. "It's hard to visualize that once there was an ice pack a mile high that went all the way to Iowa. It took that much ice to make the glacial scars in the granite."

I probably won't see much glacier scarring on Oak Island. Unlike most of the islands in the lake, it has retained a sandy shoreline. And Oak Island Resort, when I arrive, comfortably matches my picture of a midwestern fishing camp: cozy, rough-hewn log cabins (painted the dull fire-engine red that is standard for Minnesota cabins); no telephones or television; a large fish-cleaning shed; and a rec room complete with old magazines, pool table, assorted exercise equipment, and fading photos of almost man-size muskies and arm-length Northerns and walleyes. There's even a hot tub of sorts—a recycled bait tank that was formerly the home of thousands of minnows, until the resort's owner, Jack Hawkins, decided this would be a better use for it.

"I'm the guy this place runs," is how Jack introduces himself. He's sporting a two-day growth and a gasoline-and-grease-stained shirt, and his eyes have the look of a man who has spent months at sea in a small open boat. He shakes hands, tells us, "The fishing's been great," and then vanishes back inside the motor shed, where the guts of scores of outboard engines hang from the ceiling.

After unpacking, I stand on the grass outside my cabin and survey the lake. Ducks are swimming lazy circles in front of the dock, blithely ignoring a beaver working on his lodge a hundred feet away. Toward the north end of the island there's a lone fisherman in a boat, casting near a flock of white pelicans.

Across the bay lies a shotgun pattern of islands, most of them in Canadian waters, as a large maple leaf marker in the middle of the channel indicates. Flag Island is immediately opposite me, as is Cyclone Island, home of Canadian Customs and a small Ojibwa Indian curio store in a secluded cove.

At the south end of Oak Island stands the Bay Store, a square, tin-sided building perched on the edge of the rocks overlooking the beach. Constant traffic streams in and out: Floatplanes drop off guests, boats come in for gas, visitors stock up on lures and 3.2 beer. Grace Webb and her husband, Bucky, own the Bay Store. They are the closest things to figures of U.S. authority for miles in any direction.

"My husband is Customs, and I'm the Post Office," says Grace, hurriedly sorting mail for the thrice-weekly pickup. Grace's parents bought the store in 1927, using it as a base from which to ship fish, blueberries, furs, and wild rice down to mainland Minnesota. She has spent most of her life on the island, coming here when she was 14 days old. Like other children, she attended the local one-room schoolhouse and learned to adjust to the isolation.

"You can be lonely anyplace," she says. "I've worked in Chicago. I get lonesome here, and I got lonesome down there. Everybody gets lonesome. Some people can deal with this way of life, and some can't."

It's not just the lack of people that the island's eight permanent residents have to contend with. The nearest hospital is 40 miles away, the nearest shopping center, 15, and the nearest police officer, 8. Despite the inconveniences, Grace still would not move back to Chicago.

"There's too much noise and too much light in the cities," she says, shaking her head. "There is a certain freedom that people give up, but you can live here."

As Grace talks, the rumble of a floatplane announces the arrival of Bob Hansen. Bob has been flying the mail in for more than 40 years now. Despite the remoteness of Oak Island, he says, mail moves quickly, arriving in Minnesota in one day. And now he's in a hurry. The pontoons of his floatplane are packed with bait—300 dozen live minnows.

Returning to the cabin, I see a boat slowly pushing across the bay toward the dock. Four sunbaked men climb out, dragging two massive coolers up to the cleaning shed.

Inside one cooler are at least eight hefty walleyes, sloshing around in a mixture of melted ice, blood, and beer.

"Got 'em all trolling," one of the men says, hanging the fish up on a rack for a picture.

"Stand underneath," he yells to a friend as he focuses his camera. "Let them drip on you!"

Tomorrow it'll be me standing there, I tell myself.

THE NEXT MORNING WE SET OUT EARLY, HEADING UP TO CENTRE ISLAND RESORT, just across the border, to buy our Canadian licenses. As we pay our ten dollars for our four-day passes, a family of mud swallows emerges from its nest under the eaves of the wood-and-stone building and swoops through the air around us.

"Rumor has it that a lot of Chicago gangsters used to come up here, people like Al Capone," says Floyd Ballinger, the resort's owner, as he stamps our passes.

"The house was built by Tony Wons, who had a syndicated radio show in Chicago in the 1930s. It's made of hand-laid stone. You can see the names in the concrete of some of the Indians who helped him. It's not winterized, so the longest we can stay is late October."

Inside the dining hall North Woods memorabilia dominate the decor: a 50-year old birch-bark canoe, trophy fish, snowshoes. But what attracts our attention is the glass case full of lures for sale.

"Lot of people been buying these," says Floyd's wife, Joanne, handing me a Lazy Ike, a three-inch piece of curved plastic painted black and white and trailing two wicked-looking triple hooks. When the Lazy Ike is trolled through the water, she explains, it dives down and wiggles like a leech. I like it. I buy four of them, all different colors. I also pick up three Rapala Shad Raps, the "Original Finnish Minnow." This is a four-inch-long lure of carved balsa wood, intricately painted to look like a tiny fish.

Legal and properly outfitted, we're finally ready to hammer 'em, as the locals say. Our first stop is Poacher's Bay, on the west side of Falcon Island. For the next six hours Bill and I track fish on the boat's radar, but our prey remains suspended in the cooler water far below our lures.

And that's it. Not a strike, not a hookup, not a nibble.

AFTER DINNER THAT NIGHT, JACK CONSOLES US FOR OUR EMPTY COOLERS. "IT'S the unseasonably hot weather," he says. But he insists the fish are there. "If you're going to be successful at fishing, you should relax," he adds. "A fish can feel tension on the line, and it doesn't like it. The most intense people are the worst fishermen."

The next day, after another morning spent fruitlessly chasing fish on the scanner, Bill and I meet four Winona State University professors who seem to have the lake figured out. They're enjoying lunch in a secluded bay on Falcon Island. Someone has placed a picnic table here, in a grove of white pine near the beach. It's an idyllic setting, and the knowledge that the woods are full of bears, timber wolves, deerflies, and wood ticks make this pocket of paradise even more enjoyable.

"We're having five walleyes for lunch," says Byron Smesrud, an engineering professor, now the cook. "It was hard catching fish in the heat."

As he talks, he flips the pan of fish, fresh fillets they'd cleaned on the rocks. In addition they've put together a typical shore meal: fried potatoes, onions, canned beans, canned fruit, and cookies.

"If this had been our first year up here, we would have struggled," says Dan Bloom, a physics professor. "But as it is we've come back with fish every day."

NOBODY GETS SKUNKED ON LAKE OF THE WOODS. AT LEAST NOT ACCORDING to Jason Grahn, a 17-year-old guide from Roseau, Minnesota, who spends summers working on Oak Island. "I remember one guy who caught a 25-pound Northern," says Jason slowly, as we sit in front of a cabin that evening, watching high clouds blow in from the west. "And his group pulled in their good-size walleyes—a real haul. Among four people they had 160 pounds of fish in one day."

"So what's the secret," I whine desperately. "What am I doing wrong? What *should* I be doing right?"

"It depends on the weather," Jason says sagely. "On a cloudy day, when it's dark, you use a bright spinner or a Lazy Ike, silver, or hammered gold, or even pink. On bright days you use darker colors—orange, purple, or perch green and gold."

With Jason's advice about lures and Jack's admonitions to relax ringing in my ears, Bill and I set off for an evening of fishing at Deepwater Bay. There, with a moose as my witness, I hook my first fish.

The next morning we forego fishing to visit Cyclone Island. Like many of the Canadian islands in the lake, Cyclone is part of an Ojibwa Indian reserve.

"There's only one family on Cyclone," says Maureen Powassin, who runs the small Indian curio shop on the island, "all Powassins. It gets kind of boring in the winter, but it's nice because it's so quiet."

Maureen speaks with a lilting English accent that sounds almost Liverpudlian. She can understand Ojibwa but doesn't speak it much.

But that doesn't mean the Ojibwa have lost all contact with their traditions. There are old pictographs carved into a rock near Bishop's Point, at the north end of the lake, and Maureen shows me the spot on the map.

"It's called Picture Rock," she says. "You splash water on the rocks, and a red picture of a bear appears."

It's too far for our little rented boat. Instead we turn back to rendezvous with Jason, who has promised us that with him as our guide, our luck will really change.

We head out toward the dozens of islands that rise out of the water on the west side of Falcon Island: Betty, Windigo, Windfall, Burnt, Bluff, High. Each island is different. One has a steep rock drop-off, another sandy beaches, a third, small weedy bays where Northerns lurk.

Jason has us remove the lures from our lines and in their place put on some of his personal Li'l Nabber leaders, with live worms or leeches. It takes a few tries to wrap a worm around the hook, but that's infinitely easier than hooking on a leech.

Leeches are one of the baits of choice here, not only for their appeal to walleyes but also for their toughness. Leeches seem to have a primeval tenacity; they last forever in the water, enduring strikes, heat, and multiple hookings. The lake is full of them, and they're not dangerous. But I still find them creepy to handle. When I touch them, they writhe and blindly search for a bit of soft skin to fasten their suckers into.

"Position the head so the sucker will grab on to your thumbnail," Bill tells me. "Then slip the hook through the back."

"Oooeee!" yells Jason, trolling with one hand and steering the boat with the other. "I've got a hog on!"

Within minutes, everybody on board has a hookup, and for a frantic hour we're stumbling over each other in a fishing frenzy, tangling lines, fighting over fat leeches. We throw back more than a dozen perch but hang on to a half-dozen bass and eight walleyes—nothing in the trophy range but definitely keepers.

At noon we motor into a secluded bay on Drift Island. While I build a fire, Jason fillets the walleye on top of the cooler. As lunch cooks, I wander to the edge of the lake and watch a loon nervously scurry around her nest, mindful of an eagle in a lightning-ravaged pine nearby.

All afternoon we hammer 'em, nail 'em, kill 'em, whale on 'em. Splattered with fish blood and blind to deerflies that swarm around us, we cast and reel, cast and reel, switching lures in record time, patching together leeches and nightcrawlers with the agility and speed of an assembly line of Japanese integrated circuit assemblers. The fish seem to respond to our madness, hitting on just about anything we throw into the water.

By the time we return to Oak Island at twilight, we're all fished out. The cooler is jammed with Northerns, walleyes, and bass, and when we walk up the dock, it's our turn to bask in the envious stares, to answer the surly questions about where we've been. As the sun edges over the mainland, we hang our fish on the rack, arguing happily about who caught which. And when it's my turn to pose, I bend down, swat away a mosquito, and, yes, let them drip all over me. [1989]

A Time in Newfoundland

CAROL McCABE

SWEATERS AS THICK AS hooked rugs are stacked on the shelves of a Newfoundland craft shop. The shop smells of wool, spruce, and the salt that seasons the island air from Jerry's Nose to Joe Batt's Arm. Two gleeful shoppers insert themselves into pullovers substantial enough to see a fisherman through a patch of the cold fog the locals call mauzy weather. Others discover double-knit mittens and socks, small hand-carved wooden dories and logging sleds, and cookbooks with recipes for sturdy Newfie grub: stuffed cod, moose stew, the salt beef dish called jiggs dinner, and the steamed pudding called figgy duff that goes with it.

But to my mind, nothing else in the shop says Newfoundland as well as a jokey lapel badge, "The world ends at 12:00 . . . ," it reads, "12:30 in Newfoundland." I add one to my heap on the counter, on top of a sweater as heavy as figgy duff. One of my favorite things about Newfoundland is Newfoundland time.

This island is the smaller part of the Canadian province of Newfoundland (pronounce it "Newf'nLAND"), which also encompasses mainland Labrador. It is closer to Italy than to westernmost Canada. St. John's, its capital, is 600 miles from the nearest city, Halifax, Nova Scotia.

So far is Newfoundland from its neighbors that it's in a time zone all its own. Like other things on this island, the local time is a bit quirky. Thirty minutes, not the usual hour, is the official difference between Newfoundland time and time zones to the east and west.

Unofficially, though, the time gap between mainland North America and this wave-washed Brigadoon is more on the order of half a century than half an hour. With a Boston-size population spread across a New York State-size island, Newfoundland is as uncrowded as Thoreau's Cape Cod and as remote in time.

Nineteenth-century New England must have resembled contemporary Newfoundland, with its self-sufficient citizens, its open shores, twisting back roads, and higgledy-piggledy villages. From here the North American mainland seems a distant and unlikely place. On a Sunday morning a few days after buying the end-of-the-world badge, I find myself in Fogo, one of Newfoundland's oldest settlements. There a shopkeeper directs my attention to the date of the newspaper I've taken from her rack.

"That's last Sunday's paper, m'dear," she points out. "Today's will be in on the Tuesday. Scattered times it won't come until Wednesday. It makes no difference to the livyers, but I thought that being a come-from-away, you'd want to know."

Her speech sounds quaint to a mainlander's ear. The ancestors of most of today's Newfoundlanders arrived from England's West Country and southern Ireland in the 18th and 19th centuries. Left alone for a couple of hundred years out here in the North Atlantic, Newfoundlanders preserved the vocabulary and accents of their forebears. Not only was Newfoundland isolated from the rest of the world, but one settlement was isolated from another for want of highways. The result is that even today, people who live an hour's drive apart may sound no more alike than people from Jersey and New Jersey.

Pauline Peddle, for one, says she can't understand her own father-in-law. Pauline is the enthusiastic upstairs waitress at the King Cod restaurant in St. John's, where oval platters piled with fish cakes, fried cod, and cod tongues fried with salt pork crowd the blue-checkered tablecloths.

Pauline and I got onto the subject of accents when I had a moment's trouble understanding some directions she gave me: "Now you soaks your hard bread, which is the brewis, in water," she was saying as I scribbled the recipe for the quintessential Newfoundland dish, a rib-sticking mess of salt codfish, hard bread, onions, and salt pork called fisherman's brewis. That's pronounced "broose," I learned.

Pauline serves lashins of it. Lots, that means. She smiled agreeably, said a lot of come-from-aways have trouble with the accent. "There's some I can't make out myself," she said. "My 'usband's father talks to me, and I just nods and smiles where it seems like t'right place."

Newfoundland expressions are endlessly colorful. "Where do you belong?" a Newfie will ask a stranger, meaning "Where are you from?" A celebration is a "time," an expression also heard in parts of New England, as in "They're having a time for Athol Kearney at the Fishermen's Hall." A fellow too lazy to fish hangs around shore: Thus, a "hangashore" is a lazy person. So is a "nunny-fudger."

Many expressions have to do with fish and ships. There are the comparisons: rough as a dogfish's back, deaf as a haddock, and foolish as a caplin—a smeltlike fish that beaches itself to spawn. "Long may your big jib draw" is a circuitous way to wish someone luck. "Your tawts are too far aft" means "You're wrong." (Before you can understand that sentence, you probably need to know that "tawt" is the local version of "thwart," which is the cross seat in a boat.)

IT WAS NEARLY 1950 WHEN NEWFOUNDLAND VOTED, BY A SMALL MARGIN, TO throw in its lot with Canada. Despite confederation, mainlanders have treated the island like a kind of briny Dogpatch and aimed the usual jokes eastward: "Did you hear about the Newfie who went ice fishing? He brought home ten pounds of ice." And, "How can you get ten Newfies into a Volkswagen? Throw a codfish in first."

But now, like the Louisiana Cajuns who were also mocked for their old-fashioned ways, Newfoundlanders have come into their own. The island is Canada's maritime attic, stuffed with precious old things. Newfoundland's folk culture includes more than old crafts, curious speech, and curiouser cookery. Best of all, perhaps, is the music.

For a sample, drop into Bridgett's Pub in St. John's any Wednesday evening. That's when this cheery, oak-paneled room is taken over by the St. John's Folk Music Club, a collection of fiddlers, accordion players, tin whistle blowers, and other traditional musicians. If you're lucky, you might hear Phyllis Morrissey's lovely clear voice singing the plaintive ballad "Let Me Fish Off Cape St. Mary's":

Let me sail up Golden Bay
With my oilskins all a'streamin'
From the thunder squall—when I hauled me trawl
And my old Cape Ann a gleamin'
With my oilskins all a'streamin'. . . .

Another Wednesday regular at Bridgett's is Kelly Russell, a fiddler in his early thirties, who learned to play from Rufus Guinchard, an island legend. Kelly grew up in St. John's but loved the music of the outports. "There has been a rising consciousness in my lifetime that we have something unique and valuable in our Newfoundland music," he tells me.

More than a thousand folk songs have been catalogued in Newfoundland, some brought from Europe, many others homespun. Some of this traditional music can be heard at regional folk festivals held around the island almost every weekend in summer.

Topical songs were polished in outport kitchens, "where people would sit around and sing to pass the time," Russell says. "They'd write songs about events, anything from a goat getting into somebody's garden to a shipwreck on the shore." Events continue to inspire songs; several were written after the 1982 collapse of an oil platform on the Grand Bank.

OUTSIDE BRIDGETT'S PUB, ST. JOHN'S CARRIES ON AS A PROUD, IF LOW-KEY CAPItal, of a province. The city, whose recorded English history began when John Cabot sailed up on St. John's Day in 1497, bills itself "the oldest community in North America." St. John's is wrapped around a snug harbor, behind headlands that protect it from the wild Atlantic. That anchorage and the fish circling the shores have lured fleets from many nations for centuries. The first letter sent home from North America to Europe is said to have accompanied a load of codfish eastbound from St. John's in 1527.

The capital today retains the feeling of an English town, its downtown houses sporting wrought-iron fences, beveled glass windows, and chimney pots from which a social smoke rises on dull, cold mornings. Brightly painted wooden houses climb the hills above the harbor, their windows overlooking the movements of ships from around the world.

St. John's is small enough to explore easily and has lots to see, from the wave-beaten cliffs and restored lighthouse at Cape Spear to the craft shops of downtown Duckworth Street. The Newfoundland Museum is the island's version of the Smithsonian Institution, crammed with everything from the plans of early forts to Indian relics. There are exhibits devoted to the province's history, art, crafts, seal hunting, and, of course, to cod.

The Commissariat House was built in 1818 for the assistant commissary general of the British army. Restored and furnished, it is open to visitors. More history is evident at Signal Hill, site of the last battle of the Seven Years' War, between the British and the French, in 1762. During July and August the Signal Hill Tattoo is performed there, re-creating drills and battle formations.

St. John's is also the easternmost terminus of the Trans-Canada Highway, Newfoundland's main road, which stretches 565 miles from the west side of the island and Channel-Port aux Basques, where the overnight ferry docks from Nova Scotia. The Trans-Canada Highway was completed only about 25 years ago, sending sled dogs into retirement and car dealers into prosperity. Most of the TCH traverses the island's interior forests, a landscape of spruces and birches, barren rocks, and moors covered with spongy scrub known as tuckamore. The highway passes bouncing salmon rivers, glimmering lakes, and kettle ponds. It gives access to two national parks, Gros Morne on the west and Terra Nova on the east, and many of the island's 77 provincial parks.

Most nonurban Newfoundlanders live away from the TCH though, along 6,000 miles of coastline that's as tattered as an old fishing net. Bays, sounds, coves, reaches, bights, tickles, and arms cut the land into islands, capes, points, and noses. Down the roads of each peninsula, pink and purple lupines run riot in churchyards, cows graze in seaside pastures, and church steeples are the tallest things man has made in the villages that hug every harbor.

The island's wild west coast has the most dramatic scenery, with its fjords and heavily forested mountains that come down to the sea. Even in summer here it's not unusual to see icebergs floating just offshore. At L'Anse aux Meadows at the tip of the Great Northern Peninsula, the restored sod huts of a 1,000-year-old Viking settlement are part of a UNESCO World Heritage Site. The Southern Shore has Ferryland, where Lord Baltimore began a colony before Maryland was a gleam in his eye. Nearby is a famous seabird sanctuary. And on the eastern edge, St. John's is awash in maritime history.

BUT NO PLACE IN NEWFOUNDLAND IS LOVELIER THAN THE PENINSULAS AND ISlands on the north coast between St. John's and Gander Bay. Here are the haunts of a pirate king who commanded a navy of 5,000 Newfoundland fishermen. And here old whaling ports and fishing towns have odd names like Cupids and Dildo. Each village seems more preposterously lovely than the last. When a Newfoundlander told me, "Oh, me love, if you see Heart's Content, you'll never get it out of your eyes, " I knew what he meant.

But not even Heart's Content, maybe not even Salvage, can compare with the picture prettiness of villages on Fogo Island and its neighbor, Change Islands. Fogo's villages include one named Joe Batt's Arm, reputedly in honor of an 18th-century sailor who somehow became separated from one of his limbs there. Each is a cluster of clapboard houses shaped like children's blocks: cubes, half-cube add-ons, and low pyramids on top. The houses, many of them white with cornerboards painted red or green, have three windows on the front upstairs, two windows and a door down.

Nearby are big churches with marble headstones lost in grass and wildflowers. For almost every house, there's a fishing shed called a stage, with a dilapidated pier stretching into the harbor. Peeled twig fences like the ones at Plymouth divide small fields beside the sea.

Michael Newman stands in the sunlight outside his house, the breeze ruffling his flossy white hair. A graveled lane twists down to the shore, where buttercups bloom against an improbably blue shed with a rack of antlers above the door. His house and those of his neighbors, arranged in no apparent order, seem to grow out of the rocky coast like a patch of tough posies.

On this August Sunday in Fogo, the wind is soft and the air clean enough to drink. "There's none of that pollution that you get down in the Boston states," the old man says, using the local name for New England. "When that dirt comes off t'mainland, it would have to make a left hand turn to get to Newfoundland. It just can't make it, m'dear."

"A lovely day, me son," calls a passerby. "And on t'radio they're advertisin' it to be like this right up to T'rsday."

Because it is Sunday, fishermen's yellow oilskin aprons and pants are hanging from the clotheslines beside the houses. No fishing today. They splash color against weathered walls and fences draped with nets.

In this land of ferocious winters, every day as sweet as this one is a gift. "Come winter, t'ere'll be hice all t'way to Change Islands, and snow over tops of t'fences," Michael Newman says. Using the Anglican church steeple as a landmark, he directs me to the museum called Bleak House. I'll find much there from the old fishing days, he says. "Oh, my dear, Fogo would be nothing at all without fishing."

"In the winter, you really can't get off t'island," Shirley Heath continues, as she shows me around the museum. "Especially January, February, March, when t'ice comes. Then you stay home, unless you have to go t'ospital."

Daughter, granddaughter, and great-granddaughter of Fogo fishermen, Heath has just graduated from high school and has a summer job at this former merchant's restored home. Hanging on the wall is a list of "Rules for Reviving Persons Rescued from the Water." The final rule is, "Caution. Do not be discouraged if animation does not return in a few minutes. The patient sometimes recovers after hours of labor."

And sometimes not. Memorabilia of lost sailors and fishermen are displayed at most such museums throughout Newfoundland. Earlier I had met a woman visiting from Nova Scotia. Her father was born on Fogo, she said. Three months before his birth, his father and his uncle had gone out fishing and never come back. A few months later, their boat washed up on Change Islands. "My grandmother had lost three small children to sickness, and after Dad was born, she gave up and left Newfoundland. It was a hard life."

Days later I think of her as I sit alone in an Anglican church on a windswept hill on Change Islands. As I read prayers of thanksgiving from the *Book of Common Prayer,* I can hear the sea endlessly washing up onto the stony beach below. For centuries the fisherman's life was the Newfoundlander's lot. The cod fishery was what brought his ancestors here and what would keep food on his grandchildren's table.

Back in England after his 1497 landfall, the Venetian John Cabot reported, "The sea there is swarming with fish, which can be taken not only with the net, but in baskets let down with a stone." But by then Cabot's findings were older news than the baseball scores in the paper in Fogo. Before the Venetian "discovered" Newfoundland, fishermen from several nations had come seeking food for the kitchens of Europe.

For nearly 500 years after Cabot, the supply of fish seemed inexhaustible, as was the number of men willing to go to sea in open boats after the big cod. William Godwin, Shirley Heath's great-grandfather, remembers rowing five miles in an open boat to his fishing ground. After the rowing, there was the hauling, a task described by retired fisherman Lawrence Gibbons in a wonderful book called *This Marvellous Terrible Place.* Gibbons tells about jigging cod in shoal water: "A steel hook, stuck into a big fish. Twenty pounds of wild fish, just five fathoms down. You haul in 150 of them, your hands wet and line slipping on your fingers. By and by, the skin comes off, and then the salt water eats away your flesh. . . . You talk about sore. Oh, blessed Virgin who's dead. Oh, misery."

"Today fishing's a picnic compared to them days," Godwin says when we talk at his kitchen table in Little Harbor. "Get in one of those speedboats, push a button, and you're gone."

He says he regrets just one thing about his life. "They didn't have a school where I was a boy, and I went fishing young. I never got an education." Still, he says, "I never took the Lord's name in vain but five times in my life, and I never spoke rough with a woman."

I want to ask him about those five times, but he is a bit deaf and, at age 99, tired. It is time to let him rest. His great-granddaughter will have a different life. When the summer is over, she says, she'll go off the island to college. She wants to work in hotel management. Of 69 students who graduated from Fogo's high school this year, 50 will leave the island for college or technical school. "You can't depend on fishing any more," the girl says. "At the beginning of this summer there was no fish at all. No fisherman would encourage his son to go into fishing now."

Local newspapers explain why the grandchildren of fishermen are leaving Newfoundland. "Frustrated by depleted fish stocks, dozens of Fogo residents are moving to mainland Canada in the largest exodus islanders have seen in 20 years," a St. John's paper reports. The story quotes Paddy Shea, 23, who says, "Times are tough in the fishery here and I don't see them getting better." Sean Ford, who's 21, agrees. "Every year the fishery gets worse. This year what fish I'm catching, I could pick up on the beach. . . . "

Newfoundlanders blame the fishing methods of the European Economic Community—Spain and Portugal in particular—for overfishing. The Canadian government is using diplomacy in an attempt to resolve the problems with the EEC and to rebuild fish stocks, but no one foresees a quick solution. More oilskins will stream from the clothesline.

The young people who leave will come back often. They return because nobody in Toronto or Winnipeg knows how to fry cod tongues or make a decent jiggs dinner. Living upalong, they can't smell the salt fog or see roses, lilacs, daisies, daffodils, tulips, and lupines all blooming at once. They'll come back to hear the seabirds squall and see moose lope along the highway. They'll long for roads whose shoulders are always swathed in fishnets left there to dry. The Canadian departure lounges are full of them, flaunting their Newfie accents, adjusting their watches, talking of the "times" being planned for them back home.

ON MY LAST DAY IN NEWFOUNDLAND, I STOP ON A ROAD ABOVE THE SEA TO SNAP a picture of a black-haired girl named Ruby. Dressed in a red sweater, she is hanging laundry outside a big, square, bright green house.

"Come in my 'ouse and meet me muther," she urges. "Come on," and I follow her.

Inside are her mother, her sister, and her sister's baby. In the unwary way of Newfoundlanders, the women share their hospitality. They offer me tea and the baby to hold.

"And where do you belong?" Ruby's mother asks.

In Virginia, I tell them.

Scattered times, when I remember Newfoundland, I think I was foolish as a caplin. [1990]

Nantucket: New England Spirit Distilled by the Sea

JAN MORRIS

SOME ISLANDS ARE UNDENIABLY more insular than others. Few seem more absolutely surrounded by water than Nantucket on a gusty night of the fall, when the wind blows wild off the open Atlantic, and a whiplash rain assaults the waterfront. Then, as the big car ferry from Hyannis cautiously eases itself alongside Steamboat Wharf, and the lights of Nantucket town gleam wetly through the downpour, it feels as though you are arriving somewhere infinitely remote and oceanic, barricaded against all the world by the stormy sea itself.

The ship looms white and humming above the quay. The trucks and cars come rumbling off. There is a gleam of oilskins, a mingled smell of rain, gasoline, salt, and wood smoke from the town. Backpackers stumble off into the night. Pickups are loaded with children and junk. People greet each other as though they have traveled not just the 20 miles from Cape Cod but a couple of thousand miles from another continent. There is an unmistakable feeling of adventure to the landfall, and no doubting for a moment that this is an island disembarkation. The very air is island air, the shouts in the darkness are island shouts, and the sensations of the sea are all around.

Romantically though I present them, these impressions are just. Nantucket really is islandness epitomized, not just physically but historically, philosophically, emotionally, socially, and aesthetically too. Its early settlers included a predominance of Quakers, anxious to put a safe distance between themselves and Massachusetts's mainland Puritans (who, in their godly way, detested them); it has lived defiantly ever since, generally bucking trends and ignoring odds to preserve its own famously individual character. Native islanders are extremely proud of being different. They feed upon their own extraordinary circumstances, long since flavored by rich additives of legend and anecdote, and they are engaged now as always in an interminable battle to keep Nantucket as thoroughly Nantuckety as it possibly can be.

In this too the island is allegorically insular: For among all geographical kinds, beautiful islands are the most vulnerable to exploitation. Keeping an honorable equilibrium between past and present, between real life and pretense, is the classic challenge facing all such communities today, and nowhere is it taken up more hardily than in Nantucket. For myself I could not help feeling, as I shouldered my own bags on Steamboat Wharf and set off into town, that I was entering some beleaguered sea fortress; so bravely shone the floodlit towers through the rain, so ruggedly cobbled was the street beneath my feet, and so pugnaciously tight shuttered seemed the glorious old buildings between which I passed damp, wind-blasted, and leaf-scuffling to my inn.

ON AN EMPTY HEATH, THE VERY NEXT MORNING, I STOOD IN BRILLIANT SUNshine surveying the entire island. The storm had passed, the wind had dropped, last night's feeling of civic resolution had given way to a lovely calm. Nantucket is about 14 miles long at its longest point, 3½ miles wide at its widest, and most of it is scrubby heath known here as moorlands. I stood beside a little lake that morning, a thicket of small pines behind my back, and the heath around me was partly dry grassland, like pampas, and partly a prickly wilderness of berry bushes and small scrub oak a foot or two high. The air was brilliantly clear, the silence complete. I felt utterly alone, except for the twitchy birds that, moving low and nervously, as though they were flying below the winds, hopped and scuttled from one patch of scrub to the next.

My horizon was bounded everywhere by the sea, and it was easy to imagine that aeons of ocean winds, perpetually scouring the island, really did keep the birds flying low and the foliage stunted. Nantucket is a horizontal island, nowhere much higher than 100 feet above sea level, and it keeps its secrets hidden in almost imperceptible folds of the landscape, so that when I drove away from the heath I repeatedly discovered surprises.

Across one rise, I found a wide tranquil pond with swans, geese, and seagulls loitering on it. Across another, and suddenly there was a sea creek bobbing with boats. There are little hidden forests on Nantucket, and herds of deer, and unsuspected patches of heather. What's this stone fountain at the road's edge, far out in the countryside? Why, it's a memorial to Ben Franklin's mother, Abiah Folger, who was born nearby. Why is this track so unexpectedly well defined? Because it leads to one of the world's largest contiguous cranberry bogs, a great patch of sludge, scrub, and ditching altogether out of sight of the highway.

This is anything but a desert island—you are seldom out of sight of a house, even up there on the moors—but around the edges it has its solitary situations, of the true American kind: long empty beaches with rickety fences and lighthouses, tumbled grassy dunes, or the long sea spit called Coatue, which curves all around the north shore of the island like a sandy rampart and is a gloriously ornithological, scallopy, four-wheel-drive sort of place, with the statutory lighthouse at one end of it, and one or two dilapidated shacks built by fisherfolk long ago.

I could happily putter about this desultory landscape for weeks, watching the birds, looking out for those deer, or not very hopefully searching for the little salamanders that are also alleged to frequent it. Sooner or later, though, the twin towers of Nantucket town, or the summoning of its bells through the silence, would draw me into town. If country Nantucket is laid-back, patternless, free and easy, urban Nantucket is all decision, all trim and compact order, all of a piece. It is also, as it miraculously happens, perhaps the most exquisitely beautiful little town in North America: a true prodigy of history and architecture—"shut up, belted about," as Herman Melville said in the most frequently quoted of all Nantucket quotations, "every way inclosed, surrounded, and made an utter island of by the ocean."

The beauty of Nantucket town, which has been repeatedly knocked about by destiny, is essentially *sturdy*. Its homes range from mere shingled cabins to town houses of opulent splendor, but they are nearly all tough and self-reliant structures—island structures, in fact. Reliability rather than grace is their hallmark, and they stand there, street after street, in attitudes of neighborly but hardly gushing resolution.

Street after *crooked* street, because although the town has few straggly outskirts, its shape is wonderfully intricate. Around the nexus of its central square (which is just a wider part of Main Street, really) webs of roads and alleys wander away, sometimes briefly falling into symmetry, more often ambling amiably from one intersection to the next, doubling back upon themselves or petering out in bosky culs-de-sac.

It follows that texture rather than form is the chief attraction of this architecturally exemplary seaport. The junction of clapboard, brick, and shingle, the tangle of green foliage that softens every vista, the elaboration of tall brick chimneys, external staircases, porches, roof walks, and verandas, the rich variety of fenestration, the ironwork, the lampposts, the horse-head hitching posts, the multitude of odd conceits like marble slabs, marker stones, memorial plaques, and hanging signs—all this delightful complication is matched by the patterning of the ground itself, which is a mosaic of cobblestones, slabs, and brick paving, all fertilized, when the fall winds blow, by the scattered rotting leaves.

Surprises abound here, and serendipitous cameos. A tabby cat, sitting bolt upright outside a tall white house, looks as quaintly out of scale as a cat in a naif painting. A woman in a hardware store carries a macaw upon her shoulder. A boy plays the flute on Main Street, and when you go to get some cash at the Pacific National Bank of Nantucket (established 1804), they give you an anecdotal checkbook; each "check" tells a quaint tale of Nantucket, with jolly pictures too.

I know of nowhere more beguiling simply to perambulate, guided by those two towers—the one of South Church (which is gilded in 22-carat gold, against the climate), the other of North Church (whose steeple was hoisted up there by helicopter, after 140 years without one). No traffic lights will delay you as you stroll—there are none in Nantucket—and even the pedestrian crossings are fun to use, since they often consist of stone flags embedded in the cobbles, like stepping-stones across a stream.

Downtown Nantucket is in splendid physical condition, and hardly a house is not worth looking at. There are the grand mansions of 19th-century shipowners, gleaming white and portico'd in Greek Revival style, urbanely brick and cupola'd in Colonial Classical. There are the foursquare sensible houses of sea captains, parading in a comradely way down Orange Street (where 100 shipmasters are said to have once lived all at the same time). There are the shingled cabins originally inhabited by black seamen, recruited from Cape Verde to man Nantucket's ships. Then there is a not very terrible antique jail, with a fireplace in one of its cells and more humane toilet arrangements than they have in British prisons to this day, and Swain's Old Mill, which still grinds cornmeal with heroic creakings and groanings of its mechanisms, and the tremendously classical Atheneum library, and sundry churches, meeting-houses, and civic institutions that look precisely as they must have looked a century or more ago.

Much of Nantucket town was destroyed in a fire in 1846, but it was quickly rebuilt, and today there are 800 buildings that predate the American Civil War. The beauty of the place is no more astonishing than the fact of its survival, not as a reconstruction or a pastiche, but just as solid, just as genuine as it was when it was all real.

FOR NO, IT IS NOT ALL EXACTLY REAL. ITS BUILDINGS ARE REAL, BUT MUCH OF its life is a sort of sham. The smell I like best in Nantucket is the smell of wood smoke I smelled that first windy night; the smell I like least is that dread odor of the tourist age, the gift shop scent, compounded of perfumed candles, soaps, and ribboned sachets—the smell of Collectibles, and bric-a-brac emporia, and all the tinsel merchandising that mass tourism fosters everywhere in the world.

For a century the main function of Nantucket has been looking after visitors of one kind or another. It entered history in the 17th century as an agricultural island, hunted and cultivated by the now-vanished Nantucket Indians, farmed by the first English settlers, who were greatly outnumbered by their own sheep. Then in an epic few decades of enterprise it became the whaling capital of the world, its sea hunters extending their range from the offshore Atlantic whaling grounds to the most distant reaches of the oceans, until the name of Nantucket was associated everywhere with bold seamanship and initiative. Nantucket navigators knew the Pacific Ocean better than anyone else (which is why the bank is called Pacific National). They were chartered Nantucket ships from which the tea was thrown overboard at the Boston Tea Party. And it was a Nantucket ship that first took the Stars and Stripes up the Thames to London.

Nantucket got enormously rich on whales, and almost every facet of island life was bound up with the industry. When whaling fell into decline in the mid-19th century, the place was half-ruined. The islanders never quite recovered their terrific old flair, and by 1895 a local writer was lamenting that while Nantucket salt had certainly not lost all its savor, "the old pungency is somewhat abated by modern admixtures." He was thinking, even then, of tourism, for it was pleasure travel, if of a particularly rarified kind, that had by then rescued the island's economy.

Steamship travel had introduced rich people to the delights of the Nantucket summer and enabled them to build upon its shores the ample Victorian holiday retreats that still stand, memorials of a roomier era, encouched in lawns and attended by dainty gazebos around the island. Perhaps, as the man said, they did somewhat abate the pungency, but they scarcely impinged upon the island's hereditary arrangements. A new patrician class adopted Nantucket in the summer months, helpfully, bringing heaps of money with it, but it did not supplant the dynasties of Coffins, Folgers, Macys, Starbucks, and Swains who had for so many generations set the style of the place.

Today the admixture is less innocuous. That small company of cultivated estivants has been succeeded by a far less homogeneous multitude of summer visitors, whose annual assault upon the island sensibilities takes its breath away in season, and leaves it with severe withdrawal symptoms when winter comes. More than half those delectable homes of Nantucket town are left empty out of season; all across the island, houses and condominiums are left locked and boarded from fall to spring, haunted all too often by the cats their owners leave behind. And I am told that when summer returns, and Nantucket's permanent population of some 7,000 is swollen to its seasonal peak of 40,000, permanent gridlock all but seizes the dear old streets of the town.

All the threats that threaten islands are here in epitome. All over the countryside new houses are sprouting, nibbling away at the open moors, dutifully clad in clapboard or shingle but often grossly oblivious to the nature of their land. The first suggestion of a shopping mall has appeared, a cloud no bigger than a man's hand, on the outskirts of Nantucket town. The development companies, the real estate men, the insurance people, the gift shoppe entrepreneurs, the vendors of commercial T-shirts, the sellers of scented candles and wooden Latin American parrots have so proliferated that few indeed of the old island stores are left. Most of the waterfront area, now given over to yachts and such, belongs to a big mainland corporation, and a proposed new wharfside development threatens to alter the whole aspect of the port.

"Nice thing about a small town," I heard one man say to another on Centre Street one day, "you can do business in the street like this." But I thought he was either a corporation man himself or was whistling nostalgically in the dark, for there does not seem to me much down-home, small-town, grass-roots feeling to Nantucket commerce anymore, and the cap-a-pie feeling I sensed on Steamboat Wharf that first night, and came increasingly to admire as I got to know the town better—that up-in-arms attitude stems from the resolution of true-blue Nantucketers, whether natives or summer persons, to stop the rot, or at least keep the balance.

NANTUCKET PUBLIC MANNERS INCLINE TO THE OFFHAND, EDGING AWAY SOMEtimes into the surly. Service is by no means always with a smile. Being of a charitable disposition, I attribute this to historical osmosis and like to think it just another symptom of Nantucket's ornery island pride—part of this citizenry's stubborn resistance to everything that might taint or trivialize their island.

For the resistance is inescapable. Hardly a Nantucket conversation does not revert, in the end, to the matter of development and conservation, and hardly an issue is not entangled in controversy, from sewage to fire precautions (the town's fire chief waved to me with particular cheerfulness when we passed in the street one day, and well he might, for he had just been cleared of accepting improper favors from one of those property developers). Not much slips by in this island, and your wiliest speculator will not find it easy to evade the island regulations. The whole of Nantucket town has been declared a National Historic District, and tight rules control the look of buildings, if not always the numbers. About the only things allowed to jar with the venerable presence of the town are the rickety old telegraph poles that still line the streets, and I get the feeling that even they are almost certainly doomed.

A host of institutions, public and private, stands up for the island integrity: the historical association, which has been hard at it for more than 90 years; the conservation foundation, which owns the great cranberry bog; the land bank commission, which administers a tax devoted to the preservation of country areas; the preservation institute, which is compiling a register of every single old building in Nantucket. A third of the island terrain is now in the hands of conservation groups, and a popular fender slogan is NO MOOR HOUSES (but then I LOVE WHALES is ironically another).

Sometimes it all feels a bit too much. For myself I find the suavely lettered NO PARKING signs a little *over*-fastidious. Occasionally I pine for a slab of vulgar concrete or a gaudy poster on a wall. I would love a roistering pub or two on Main Street, and I shall regret the passing of those homely old telegraph poles. There is an *anti*-conservation movement too, called People for Nantucket Inc., and one 86-year-old islander, complaining about the strictness of building codes, told me she was thinking of emigrating to Gorbachev's Soviet Union, where she might be more free to do what she liked with her own property.

But she was joking, for she was really Old Nantucket through and through, one of the many women who are among the island's doughtiest champions. In the whaling days the women left at home were obliged to run Nantucket, and in many ways they are still the keepers of its conscience and its functions—the chairman of the Board of Selectmen, the executive director and the president of the Chamber of Commerce, the editor and the publisher of the *Inquirer and Mirror* (established 1821), and the manager of the airport are women. And it was mostly women who guided me around the defense works of the resistance.

They took me to Edouard Stackpole, for instance, the eminent historian of the whaling industry, who was deep in books and papers at the Peter Folger Museum building. They introduced me to Charles Sayle, Sr., the celebrated carver and ship modeler, who went to sea himself in sailing ships, and who works in a tumultuous treasure trove of ship plans and nautical memorabilia. They sent me out to the Bartlett's Ocean View Farm, where two cows podgily greeted me; for generations the Bartletts have been producing fresh vegetables for the island, and when their produce truck sets out its trays on Main Street each weekday morning, it parks next to a memorial to John H. ("June") Bartlett, Jr., clan patriarch, who died in 1959 and is portrayed on the memorial surrounded by animals and vegetables above the lines:

He loved the dawn—
He loved the soil—
He loved mankind—
And all Nantucket loved "June."

Reality holds its own in Nantucket, if only just. There are people farming oysters here. There are people dredging scallops in the bay. There is a working astronomical observatory (its astronomer a woman, of course). Out on the Bartlett Farm Road, Dean and Melissa Long are making the first commercial wine ever made from Nantucket grapes. Writers and artists of all categories inhabit the island, from wishy-washy watercolorists to internationally famous authors. And if the past is a preoccupation of Nantucket, it is also a living presence, if only because at the core of the island life descendants of the earliest settlers thrive to this day.

The miller at Old Mill, for instance, hitching his jeep to the turning wheel as his predecessors linked their teams of horses, is Richard P. Swain, two of whose ancestors were among the original ten proprietary families of Nantucket. The Pacific Club is still controlled by successors of the 24 men, almost all retired ships' captains, who acquired it in the 19th century; the club rooms, with their ship pictures, their whaling mementos, their cribbage champions' names inscribed on scrimshaw, and their aged players grim over the card tables in the middle of the morning, would seem perfectly familiar to old salts of 150 years ago. One Macy may have gone to New York to make his name famous, but several others are still around, the Coffins, whose forebear Tristram Coffin was the original chief magistrate of Nantucket, are represented by 31 names in the telephone book.

In the old Friends' Burial Ground on Quaker Road, where the town edges out into the countryside, one gravestone is obscured by a small bush. I went past it one morning and made a bet with myself that after a week in Nantucket I would recognize the name upon it. I was not disappointed. Getting down on my knees to move away the prickly branches, I was gratified to find it the grave of a Folger—you know, same as the place where Dr. Stackpole works, same as the Folger Hotel, same as Ben Franklin's ma, same as Charlie Folger the furniture restorer.

All these people—well, nearly all—are captains in the fight to maintain the balance of Nantucket, and thus by extrapolation, one might say, to maintain the balance of all lovely islands everywhere. I found myself, at the end of my stay, downright proud of them—proud, too, of the condition of the island, which if half-ruined in the eyes of its own traditionalists, and fearfully threatened still, is by the standards of the rest of the Western world magically pristine.

It pays the price of its fascination, of course, in its silly tourist trappings and its somewhat self-conscious aesthetics. It is condemned, like all its peers, to permanent struggle against Philistine exploitation and degradation. But despite everything it remains truly one of the most enchanting places on the face of the earth.

EACH MORNING DURING MY TIME ON THE ISLAND I WALKED OUT INTO THE countryside before breakfast, and generally so arranged my itinerary that I reentered the town down Main Street itself. This seems to me as handsome as any urban thoroughfare anywhere, hardly half a mile long, mostly residential and unostentatious, but nevertheless a street of ceremonial magnificence. Its houses are perfectly proportioned to the thoroughfare, great American elms arch over it, and it bears itself like a structural proclamation of Nantucket's self-esteem.

In the exuberance of those autumn mornings, with the first leafy, woody, smoky fragrances of Nantucket arising, and the first stir of the town around me, I felt almost grandiloquently self-satisfied to be striding down its sidewalk; and indeed I would have strutted clean down the middle of it, like a monarch on some triumphal way, or a whaling master home from the Pacific, were it not that I would almost certainly have tripped over its cobblestones and made a very un-Nantucket fool of myself. [1990]

The Watermen of Smith Island

CARL HOFFMAN

ASK MADELINE DIZE HOW long she's lived on Smith Island, and she says, "I've been here 73 years, through tides and squalls and floods." The stories she and her husband, Dan, like to tell are turbulent and rich. There is the man who died of lockjaw from a splinter he got under his fingernail while lunging for a hog. (Dan bought the engine from the dead man's 1927 Chevrolet for five dollars and installed it on his boat.) There is the woman whose coffin rose from its grave during a flood of the low-lying island. (Her hair had grown down to her waist in the ten years after her death.) And there is the time a squall drove the waters of the Chesapeake right into the room where Madeline's mother was giving birth to one of her 11 children. (Even as friends and family implored her to move to higher ground, Madeline says proudly, her mother refused to budge.)

Smith Island—actually four small islands—lies about ten miles off Maryland's Eastern Shore in the Chesapeake Bay. Four hundred and forty-nine people live in its three towns: Ewell, Rhodes Point, and Tylerton. No palm trees sway on Smith Island; few large trees of any sort grow on its marshy expanse. It has no white-sand beaches and no mountain ranges; its highest point is five feet above sea level. Summers here are hot and humid, winters gray and cold. The island has no tillable land save the occasional garden patch. And unlike islands in distant places whose people are stranded by geography, Smith Islanders are and always have been a short distance from the thrill and opportunity of New York or

Washington, D.C. Still, since 1657 they have, like Madeline Dize's mother during the squall, refused to budge.

Smith Island was first settled by religious dissidents from St. Marys County on the bay's western shore. Today nearly all the islanders are their direct descendants; their speech still carries hints of an English accent. With a few exceptions they are named Evans, Dize, Marsh, Bradshaw, and Tyler. They survive today by harvesting crabs in the summer and oysters in the winter, just as they have for generations.

Seven days a week, at 12:30 and 5 P.M., THE *Capt. Jason*, a 42-foot fiberglass "ferry," leaves the Eastern Shore town of Crisfield for the 35-minute boat ride across Tangier Sound to Ewell. And it was as I rushed to catch the five o'clock boat, driving along U.S. Highway 13 and passing through small farming towns such as Princess Anne, that I got my first island insight: Nobody just happens to pass through Smith Island.

When the island first emerges on the horizon it is as a long, flat, gray-green sandbar. But as soon as the *Capt. Jason* began winding its way through the narrow marsh channels, called "guts," I was struck by the primal—and unexpected—beauty. Sandwiched between the blue Chesapeake and an expanse of sky was a bright, pure green strip of marsh grass teeming with life. Everywhere stood great blue herons, snowy egrets, oyster-catchers with long orange beaks, and glossy ibises. The air smelled moist and sweet, almost like new-mown hay.

After a few minutes the gut becomes the Big Thoroughfare, a narrow channel lined with Ewell's docks and crab shanties. It looks picturesque in a rough, ramshackle sort of way, with crabbing boats, stacks of crab pots made from zinc-coated chicken wire, bright fluorescent orange buoys, piles of oyster shells, and rusting car parts.

The island is four miles wide and seven miles long, much of it marsh. With the possible exception of second base on Ewell's softball diamond, there is no place where a person can stand and see dry land for more than 150 yards in all directions. This geography has created some strong borders that define the islanders, and give them their identity. First, of course, there's the island itself. It may not be one big happy family, but it's a family nonetheless, always to be defended from outsiders. Ask Madeline Dize to comment on a community meeting called to discuss a proposed public water system, and she'll look a reporter coyly in the eye for a moment before saying, "I agree with everyone 100 percent."

The towns provide a second set of borders. Ewell and Rhodes Point are separated by a mile and a half of marsh, through which a winding strip of black tarmac runs. Tylerton is accessible only by boat. Though they seem of a piece to visitors, each town has an identity.

"I'M NOT FROM AROUND HERE," FRANK DIZE WANTED ME TO KNOW AS SOON AS we shook hands over the fence separating his house from the one I was renting in Rhodes Point. "My house is up at Ewell." The one he was currently in, it turned out, belonged to his wife. He had lived there six years.

Ewell, the island's biggest town with a population of 245 people, is defined even further. I couldn't detect any topographical variations myself, but it apparently has two neighborhoods, known as "down in the field" and "over the hill."

Although technically located in Maryland's Somerset County, Smith Island is not incorporated and has no local government. Political decisions are decided in congregational gatherings. Instead of paying taxes, islanders contribute to a "community fund" that is controlled by the United Methodist Church. Money to repair streetlights comes from the same place as money to put new siding on the church. This makes Smith Island as close to a theocracy as it's possible to get in America.

"It's like having two jobs. You're the preacher *and* you're the politician," says the Rev. Kenneth Evans, who ministered to Smith Islanders until a few months ago. Until Reverend Evans arrived, there wasn't even a separate set of books for each kind of expenditure.

At the heart of Smith Island life, however, remains the blue crab. From April 1 to December 1 the search for the crustacean seems never ending. Crabbing can take two forms: laying traps, called pots, for hard-shell crabs; and dredging, called "scraping," for soft-shelled crabs. Watermen all up and down the Chesapeake lay pots; 41,855,000 pounds of hard crabs were caught in Maryland bay waters in 1988 according to the Maryland Department of Natural Resources. But only 1,143,000 pounds of the fragile softs were scraped that same year. Most of those came from Smith.

One morning I went scraping with Carroll Marsh. He looks, he says, "just like the other Carroll, the one who plays Archie Bunker." Indeed, imagine actor Carroll O'Connor with a deep tan and a pot belly, dressed in a knee-length, yellow rubber apron, rubber boots, and a baseball cap, and you've got Carroll Marsh, 54. He wasn't too optimistic about the day, so we started late; it was 5:30 A.M. Rhodes Point's streetlights were still shining brightly when we slipped away from the dock. Soon we were part of a procession of scraping boats heading out to the shoals around Smith Island. Our task was to dredge shallow beds of eelgrass in search of "peelers," crabs that were in various stages of shedding their hard shell. The crabs are taken back and placed in raised tanks full of recirculating water until they are completely soft. Then they are rushed off to markets as far away as Japan on beds of ice and seaweed.

At 6:10 Marsh lowered his five-foot-wide scrape—an oval-shaped, iron mouth holding open an eight-foot-long net—into the water for his first "lick." The boat slowed, and for the next 15 minutes we dragged the scrape in slow circles. All 17 boats worked within a 200-yard cluster. Sometimes they nearly collided. The scene reminded me of those amusement park rides in which boats circle round in a big tank of water. Thick waterman's talk crawled over diesel engines and flying spray. "Nobody's gonna make it here today!" was one of Marsh's few shouts that I understood.

At the end of each lick Marsh dumped the net's contents out in a shallow plywood box mounted on the boat's side and sifted through a cross section of bay life: gar, toadfish, stinging nettles, eelgrass, but not too many crabs. Most were below the two-inch minimum, and he tossed them overboard. Two crabs were attached, one on top of the other. "That's a jimmy and his wife," he explained. Another crab was "red: she'll shed within 24 hours." He tossed it into one of two water-filled plastic trash cans. The next was a "buster: she'll shed within the hour." It splashed into the other can. They all looked the same to me.

The lick was, however, disappointing—only eight crabs total. Marsh decided to head back. All but a few of the other watermen did the same.

The work can be repetitive, and the day exhaustingly long. If more crabs had been anticipated, Marsh might have left home at 3 A.M. and scraped well into the afternoon. Yet Marsh is part of a continuum. His father had been a waterman, and out with us that day in other boats had been a son and a brother.

"It's a good life, but it's a hard life," Marsh says. "You got your independence, and you don't have to work for nobody." There are no time clocks to punch, no traffic to fight. The 17 scrapers on the water are 17 proprietors, each waterman exercising exclusive control over his own work conditions. On average, each could be richer and more secure banging nails in Washington, D.C.'s, booming construction market. But their lives would not be their own.

Crab prices fluctuate wildly throughout a season, however, depending on supply and demand. The day I went out with Marsh, a half bushel of soft-shells—about 100 crabs—was bringing $100. It was a good price, but then demand was high because the upcoming Fourth of July weekend fell just at the time when crabs were scarce. Watermen seemed to think the season was "fair," not great but not terrible either. Yet the state of Maryland was thinking differently and for the first time was debating a limit on the number of crabs a waterman can catch.

Crab and oyster populations have always been erratic, and a nation that has sent men to the moon still can't pin down with certainty the special combination of water temperature, salinity, and currents that make a jimmy and his wife happy. Every year the mid-Atlantic megalopolis grows. And every year there are more pleasure boaters and sport-fishermen. Already pollution, disease, and overfishing have nearly destroyed the bay's once huge oyster population, which has severely hurt the island's winter income.

Smith Islanders feel the pressure. Since 1983 their numbers have shrunk from 650 to today's 449. One quarter of those remaining are 65 or older. But it's a moot question how much of the population flight is due to fewer crabs and oysters to catch, and how much results from a quest for material possessions and a fast life unavailable on the island.

The water still holds enough promise for some. Jerry Smith, Madeline Dize's 24-year-old son, recently invested $11,000 in a brand-new fiberglass boat. He sees the situation almost like an industry shakeout. "The only (islanders) who leave," he says, "are the ones who can't hack it." And Marsh points out that catches have always been cyclical. "Back in '55 the scientists said, 'This is it; it's the end. They'll never be no oysters again.' The old-timers just shook their heads. They'll be back. They always are."

One evening Frank Dize took me to a meeting in the Ewell church to discuss a state proposal to create a "public water system"—which meant essentially changing from wells to a central system, including water meters, for every home. Even before we arrived Frank said, "I think it'll be defeated."

An islander duly read out the proposal to the 50 people assembled: The system would create an "abundance" of water for new homes and businesses, fire hydrants for the fire department, and maybe even a new public swimming pool. At the start Maryland would pay for everything, except for connection charges. Then each islander would be billed according to individual meters. Before long a clamor arose. "They'll raise the prices before they even start!" shouted someone. "Swimming pools are a health hazard!" yelled another.

A vote was taken. All those in favor of the project were asked to stand. No one stood. All those against the project were asked to stand. Everyone stood.

A second issue was discussed. The old and ill on the island had to be sent to nursing homes on the mainland. Was anyone in favor of turning the vacant doctor's home into an elderly care center, a place where doctors and nurses from the mainland could come regularly? The idea touched off as much fervor as the water proposal had.

"Once an elderly person is in a nursing home on the mainland," someone called out, "he never sees his island again. That's the truth. How many have gone and never come back?"

"I know some people who've been over there nine years with perfectly good minds. If they'd been here, people could drop by and see 'em," shouted another. "Once you leave, you never come back."

And again a vote was taken: Should the idea be explored further? All those in favor were asked to stand. In unison everybody stood.

Then I heard someone say, "Once they take you, that's the end of you. You'll never see this island again." [1990]

Out of the Past: A Personal Portrait of Key West

BARRY GIFFORD

MY MOTHER AND I LIVED for a time in the house my grandmother owned in Miami Beach, before Miami Beach became what it became, and then we lived in the old Delmonico Hotel after my grandmother sold the house. In 1954, when I was seven years old, my mother remarried and we moved to Key West, Cayo Hueso (Bone Island) as the Spanish named it, where we lived in a white hotel on the beach. Because of the heat, white was the predominant shade—the sand, the glare, the clothing. The hotel was large but intimate in that there were never many guests staying in it at any one time. In those days Key West, the southernmost point in the continental United States, was not considered a tourist's paradise. Four or five thousand persons lived on the island year-round, Cubans and Conchs (descendants of the original settlers of Key West); a few tourists came for the deep-sea fishing.

I had the run of the hotel. Whenever I caught a good eating fish, a king or a grouper, I'd wrap it in newspaper, take it to the kitchen and give it to the chef, who would prepare it for me, my mother, and her husband for lunch or dinner. I'd already caught quite a few large fish by the time I was seven—my first was a barracuda off Miami when I was five. Kingfish was best, filleted with just the right amount of salt, butter and garlic.

There was a canopied swing on the beach set between a pair of high coconut palms whose trunks curved like Lena Horne's body in the picture on the album cover of the musical *Jamaica.* The hotel pool was a fenced-in area of the Atlantic. Every morning before the pool was opened for guests, the lifeguard would swim underwater and check the wire cage that was the swimming boundary to see that no sharks or barracuda or jellyfish had poked their way through during the night. One of my favorite pastimes was walking along the beach, popping men-of-war with a board and watching the blue juice spurt out over the white sand.

One evening a Conch was fishing off the long pier behind the hotel and hooked an octopus, which somehow had maneuvered beneath the pier and caught the line in such a fashion that the only way the octopus could be brought up was directly through the planks. I stood and watched while another Conch got a saw and cut a hole in the middle of the pier large enough for his partner to bring the octopus onto the wharf. Then he nailed the section back down.

A wonderful place in Key West was the aquarium, which had huge sea turtles, sharks and a giant jewfish within touching range. The aquarium is still there, though it's been expanded. Admission when I was a boy was a quarter, but the Seminole Indian who took the money let me in for free because I came every day and stayed longer than even those who came for the first time. That aquarium, which is located right on the water, and the reptile farm that was in St. Augustine were my favorite Florida places. My mother and I went many times to see the alligator wrestling. I loved to watch the gators thrown on their backs and the Indian boys rub their bellies with sand until they'd fall asleep stretched out with their tiny legs up in the air and powerful long tails absolutely still. Awakened with a slap, they'd twist over violently and hiss and slide into the shallow pool provided for them.

Days in Key West then were quiet and hot, but with that soothing tropical breeze I always used to imagine coming from Cuba, only 90 miles away. Radio Havana blared from porches along Margaret Street and from the storefronts on White Street, and at sunset the water glistened pink under a translucent sky.

I began going back to the Keys in the mid-seventies. I hadn't been there for 20 years when my brother picked me up at the Miami airport and we drove down past Key Largo—where there's still a big sign on a beachfront café shack called the Caribbean Club that says "Site of the Filming of the Famous Movie Key Largo," or something like that—to a little restaurant on the gulf called the Sea Breeze. Unfortunately, the place has been torn down now and replaced by a fast food restaurant, but it used to be the place where men met other men's wives and women met other women's husbands. The tables were covered by red-checkered tablecloths and there was a magnificent four-foot-high black walnut bar that some guy bought for his house when the place was destroyed. The tunes on the jukebox, a big original Rock-ola, were all from the forties and early fifties.

IT WAS ERNEST HEMINGWAY, OF COURSE, WHO MADE KEY WEST FAMOUS IN OUR time. He lived there off and on for 12 years in the thirties and forties. He frequented Sloppy Joe's and Captain Tony's bars, both of which still exist. John Audubon had a house in Key West, too, and it's a museum now, as is Hemingway's old place. A number of writers have made their homes in or around Key West. Tennessee Williams still has a place on Duncan Street, has since 1941, and the Florida Keys Community College even has a Tennessee Williams Art Center on Stock Island. But it's the Hemingway legend that persists and remains a steady draw for those curious literati. What captivated Hemingway about Key West, other than the isolation, the writing, the fishing, exercise and sport? Nothing else, of course. That was enough, and for some it still is.

Hemingway's shadow lies heavy on the horizon in those parts. In 1975 my Uncle Buck and I went to Nassau to crew a friend's yawl back across the Gulf Stream to Marathon Key. It was February and, though the light was intense, the temperature and humidity were moderate, making the day comfortable. We stopped at Bimini to fuel the emergency outboard and take on fresh water before continuing across the stream.

Two old black men were sitting on the jetty playing checkers and I stood and watched them while Buck and the skipper went up to the bar with the plaque of Adam Clayton Powell on the outside wall.

"Hello," said one of the men to me. "How you today?"

"Fine," I said.

"I am Uncle Jim," said the man, "and this," he pointed to his opponent, "is Doctor Jim."

Doctor Jim, a wrinkled, emaciated little man with a cotton swab of fuzzy hair, smiled and nodded.

"You in on the Morgan boat," said Uncle Jim.

"That's right," I answered. "How did you know it was a Morgan?"

Uncle Jim laughed, "Oh, I know most the boats. Them Morgans is mighty good."

Doctor Jim made a move and Uncle Jim nodded, smiling his yellow teeth and yellow eyes before executing a double jump.

"Crown me, papa, crown me!" he shouted.

Doctor Jim got up, picked his cane off the bench, nodded to me and hobbled away.

"You tired fish these days!" Uncle Jim hollered after him.

Uncle Jim stood up. He was over six feet tall but his stooped shoulders made him shorter. I figured he was close to 70 years old.

"Bet you think how old I am," he said. "Ha, ha. I am 82. July. Still strong. I know Hemingway here. Come from Key West. We box on dock. He knock me out. You know Hemingway?"

"I know of him," I said. "His books."

"Ha, ha. That some fella Hemingway. Haven't seen him in many years."

"He's dead," I told Uncle Jim. "He shot himself in 1961."

Uncle Jim nodded. "Good reason why I no longer see him. I not really been waiting anyhow."

DESPITE THE FACT THAT THERE WERE NOW 30,000 PEOPLE OR SO LIVING IN KEY West, despite its having become a far more sophisticated place than it was when I'd lived there as a boy, for the old-time locals, life in Key West was still cheap. A charter captain who'd fished with Jimmy Hoffa for 30 years told me there were folks who thought Hoffa was buried somewhere in the Gulf just off the Keys. He knew people who'd been searching out there.

"About a year before he disappeared," the captain said, "Hoffa and his wife were down here. We were going out and Hoffa decided he needed a hat or something and went across the road there to the Eagle store to get it. His wife got nervous something awful when he didn't come right back. She knew they were after him. He was trying to get his union back, you know. He came back all right and we went out that day, but I'll never forget her face while he was gone. She was a scared lady, I tell you."

This particular charter boat captain had been over to Mariel for the Cuban boatlift, along with most of the others in Key West. He was lucky, though; his boat hadn't been red tagged for carrying illegal passengers.

"Hell, we went over there with a list of relatives to bring back," he told me, "but the Cubans made us take whoever they said to take. We didn't have any control over what happened."

I said I'd seen that yard full of confiscated boats in the Navy compound.

"They'd better take care of 'em pretty soon," he said, "before they're all vandalized and they ain't worth gettin' back. The charters are having a tough enough time getting along as it is. If it weren't for the hotels hustling up parties at cut rates, nobody'd be doing any business."

I mentioned that I'd heard about the new, enlarged water pipeline that was being built from Miami, and I'd noticed the highway was being expanded.

"Yeah, pretty soon they'll have all of Miami down here. Kind of gives you the feeling this place is on its last legs, don't it?"

Like it or not, Key West is definitely not the same as it was when I lived there in the fifties. The influx from New York and New Orleans began about 1975, and gave Key West its reputation as the "Gay Riviera." New York-style discos, restaurants and clothing stores have dominated Duval Street for the last three or four years. There's even a Ramon Navarro Apartments, named after the silent film star who was murdered in Los Angeles a few years ago, and a Margaret Truman Launderette, in honor of the former president's daughter (the Trumans had a vacation home on the island). No matter how the Yankees try to spiff it up, however, Key West is still the end of the line, seedy as ever, with rummies sleeping in abandoned shacks on Caroline Street.

For me, the weather and Cuban food are the best part of life in Key West. As I've grown older, I enjoy the heat more—not even the extreme humidity of the summer months fazes me much. If it gets too hot I just lie down and wait until the evening breeze begins to blow. The Fourth of July Restaurant on White Street is the local headquarters of the anti-Castro partisans, and still sells *ropa vieja*, used clothing, for the best price. The memorial to the sailors killed aboard the battleship *Maine*, in Cordenas, Cuba in 1898, is still standing in the Key West cemetery; there is also one to the various martyrs of nineteenth-century Cuba—a piece of the *tamarindo* tree, under which the revolutionaries sat to concoct their plans in 1868, is preserved in a plastic case for public viewing.

Despite what some charter boat captains say, the fishing is decent, especially for bonefish, tarpon and dolphin. A buddy and I caught several king mackerel right off the pier behind the aquarium in an hour last May. Boog Powell, the former Baltimore Orioles slugger, runs a marine supply and repair shop on Highway 1. I saw his mirror-image son Richie, a right-handed version of portsider Boog, belt three home runs in four at-bats (he hit a single the other time) in a local softball game. And George Mira, the ex-pro quarterback, owns a pizza place near the Searstown shopping center.

From July through October is hurricane season and every once in a while the island gets ripped up, though more often than not the big storms hit farther north on the Keys, around Islamorada. But even riding out a hurricane is a matter of perspective. I've been through a few and rather enjoyed them all, though I realize that had the storm hit directly, I might have felt differently.

But the real story in the Keys, Key West in particular, is not the Hemingway legend, the seediness or the weather: it's drugs. The Keys are the main connection point for the marijuana and cocaine smugglers, the *marimberos*, from the Guajira Peninsula in Columbia. Eighty percent of the marijuana smuggled into the United States is grown and shipped from there. Boats and small planes land regularly up and down the Keys, keeping the federal agents busy. In 1980, according to records of the Drug Enforcement Administration, agents in Florida seized 4,887 pounds of cocaine, 847,773 pounds of marijuana and 15.2 million doses of counterfeit Quaaludes. They also seized $42 million in cash, automobiles, boats, planes, and other assets.

The larger amount of confiscated marijuana has become a problem for Sheriff William A. Freeman of Key West, who supervises its burning at a Department of Agriculture incinerator on Fleming Key.

"The round-the-clock guards are killing me," Sheriff Freeman told the Miami *Herald*. "By the time we get rid of a batch, there's another batch coming in. We have four million dollars worth of it outside my office window right now."

Stories of yacht-raiding pirates, mainly drug runners keeping the lanes clear, in the Gulf Stream are legion. It's no longer safe to take a leisurely holiday sail to the Bahamas. And, if it's not drugs, then it's human cargo being transported: Haitians mostly and, as in Hemingway's day, Cubans.

Key West is still a strange, isolated place. Less so than when I first knew it, but it's the end of the land and as such is the receptacle for what falls to the bottom, no matter how vigorous the efforts at streamlining. Key West is one place I just can't imagine becoming a permanent part of the twenty-first century. [1982]

Caribbean

My Time in the Yucatán

BRYAN DI SALVATORE

MY WIFE AND I HAVE fetched up on the Yucatán, the plump, eastward-protruding thumb of southern Mexico that, with the comparatively delicate forefinger of Florida, forms a lazy OK sign to pinch the waters of the Caribbean and create the Gulf of Mexico.

Los Angeles is four and one-half air-hours away. New York is one hour closer than that. Miami is closer than Mexico City. Cuba is just over the horizon.

Our destinations, however, are more immediate, less fraught: three of the low-slung islands off Yucatán's northeast shore, beginning with Isla Cancún, young, leggy celebrity of an international resort, where at the moment we are still feeling exposed in our shorts and bare feet.

Our gait is awkward, unnatural as we walk along the fine, white, limestone-based sand that is warm, never hot, to the touch. Without warning, the Caribbean rises over our feet, our ankles, and wraps itself around our knees, soaking us to the waist. The water is fetally warm. Offshore it is transparent, *rinsed,* and on this clear, breezy day, deliriously blue.

Twenty years ago, Cancún—the name refers both to *la ciudad* on the peninsula proper and *la isla,* a 14-mile-long, quarter-mile-wide brushy sandbar separated from the mainland at either end by a narrow channel—was a swampy fishing village and, except to the few dozen families who called it home, without significance. None of the accounts of the early explorers mentions it. My comprehensive atlas, published in 1967, does not show it. A popular 1971 guidebook does not list it.

Cancún was *created.* In the late 1960s the Mexican economy was being crushed by its inability to pay off loans from first-world (primarily United States) banks. An exhaustive government-sponsored survey pointed to several potential tourist centers—each of them warm, sunny, remote, and seaside—that would serve as funnels for streams of liquid currency.

Cancún was the first atoll in this archipelago—which has been dubbed *gringolandia*—and is now home to 140 hotels and counting, 14,000 hotel rooms, five malls, at least a half-dozen mallettes, two discos, one 18-hole golf course, and something like one million visitors a year. It shares some of those visitors with roundish Isla de Cozumel to the south, the peppy staging area for the aquatically active; and tiny Isla Mujeres, Cancún's nearest neighbor and the relatively funky favorite of the relatively adventurous.

LA ISLA, WHERE WE CONTINUE OUR WALK ALONG THE SAND, IS COMMONLY REferred to, in a *Blade Runner*-ish way, as *la zona hotelera.* The hotels, large as stadiums, line up along the Caribbean shore, each with a demarcated (but public) beach, each with a private circle-drive entrance, each far enough from its neighbor to ensure exclusivity but close enough for a guest's evening stroll next door for the bar/restaurant/lounge equivalent of borrowing a cup of sugar.

We pass one after another of these astonishing creations, which adhere to the mandated architectural motif, Mayan Heroic: lengthy, landscaped, earth-toned wings (they have the stacked, fertile look of Balinese rice fields) with scores of privately patioed rooms set along staircased tiers. The wings meet at great, breezy, domed lobbies the size of hangars, some thatched, some tiled, many buttressed by mighty pillars. Swards of St. Augustine grass weave beyond the decking of the elaborate, protozoan-shaped swimming pools.

Our progress is slow. The beach is narrow, steep, and cluttered. A sign warns of sand slides. A retaining wall has crumbled, and workers are pouring cement from wheelbarrows into elaborate wooden molds.

"Gilberto" is the answer to my questions. Gilberto, the 1988 hurricane, ate the beach. Gilberto smashed the walls of the hotel's disco. Gilberto dissolved the landscaping and the shoreward-most buildings. Gilberto, as well as Cancún's annual 25 percent growth rate, is responsible for this beach having become a construction area.

We leave the beach for the island's sole highway and hail a cab. The driver looks closely at us and apologizes: No, he cannot give us a ride back to our hotel. Our clothes are damp and will ruin his upholstery.

A TIDY, BLONDE AMERICAN WOMAN WITH THE RED, SCRAPED SKIN OF THE undersunblocked, begins her order tentatively, the question mark suspended over the table in the busy patio restaurant.

"*Capisce?*"

The waiter, a young man wearing a T-shirt with the restaurant's name emblazoned on it, smiles and raises his eyebrows politely.

"*Capisce.*" Her voice has risen slightly and edges from the interrogative toward the declarative.

The waiter shakes his head briefly.

"*Capisce. Capisce!*"

The waiter opens his palm and tentatively taps the menu, verso, with his forefinger. With slight, impatient jerks, she nods agreement. The waiter smiles and writes quickly on his pad. A few minutes later he brings her a parfait glass full of raw shrimp marinated in lime juice along with tomato, onion, and cilantro.

"For the lady," he says, "ceviche."

THE RESTAURANT, JUSTLY RENOWNED FOR ITS SEAFOOD, IS ON A NARROW SIDE street in Cancún la ciudad, not far from other restaurants renowned for their seafood, and dozens, scores even, of modest and not-so-modest shops. Jewelry made from precious metals and semiprecious stones. Belts and wallets. Pottery. Perfume. Designer wear. Bright hammocks. Brighter blankets. Cuban cigars. Clay replicas of Mayan idols. Not Just Any Old T-shirts. Hard Rock Cafe T-shirts. Sunglasses. Bikinis. Swimming trunks. Skirts. Dresses. Baskets. Glassware. Tinware. Papier mâché. Crockery. Gucci. Guayaberas. Black coral.

Some of the shops are open-faced; others are sealed and air-conditioned. Some utilize overly enthusiastic sidewalk sound systems, and all ignore traditional siesta closing hours to cater to visitors from other countries, other continents. The shops dominate the city, certainly *el centro*, a spider web of angled streets, connecting at wild, joyfully chaotic *glorietas* (traffic circles) and enclosing triangular, quadrangular, pentagonal, and hexagonal blocks. Green-and-white cabs, trucks stacked with empty soda and beer bottles, and tradesmen's vans comprise the bulk of the heavy traffic. It's all a bit too busy and familiar, this brash boomtown.

We rush in a cab across the small bridge connecting la ciudad to la isla, and before long, things begin to calm down. Here in the suburbs, the marquees and banners and blare of commerce are banned by Fonatur, the Mexican tourist authority.

The hotels become larger, sparser. There are sounds here, but no noise. I am on the balcony connected to my hotel room, easily the fanciest such room I have ever engaged. Below and to my right is a long swimming pool, in which there are stone stools forming a bar. To my left is an undulating putting green, with tiny pennants attached to the metal hole markers. A couple, gray-haired and trim, no doubt thickly portfolioed, are poolside. She climbs out of the water and as she dries off, I can make out the large, handsome logo of the hotel on her towel. She approaches her husband, who is sleeping on a chaise longue, a book against his lap.

She begins to speak, but is afraid of wakening him, so at peace, so far from his responsibilities in Ann Arbor or Greenwich or Atlanta or The Loop or The Street. She smiles instead, and gently brushes an imaginary speck from his sunburned brow.

This is a Cancún moment.

COZUMEL, 30 MILES LONG BY 9 MILES WIDE, IS 45 MINUTES SOUTH OF CANCÚN via rattling, barely subsonic, cab and 11 miles offshore by ferry.

The dock at San Miguel, Cozumel's sole town, is packed with vendors: of pottery, sombreros, conch shells, blankets, hammocks, hotel rooms, and diving tours. The vendors are aggressive but not ominous. Polite refusals to buy are met with smiles, nods, and shrugs. When I demur to look at the wares of one man, however, he mutters, "*Maricón*," not quite under his breath.

We pass through town, a pip-squeak clone of Cancún city, to a remote and spacious hotel dominated by a wide beach. We swim and eat lunch at the hotel's open-air, thatch-roofed, shoreside restaurant. In front of us, boat after boat, thick with diving tanks, docks briefly at a small landing to unload strong large men and young fit women carrying duffel bags of diving gear. They drop their equipment and rally around the bar, sunglasses pushed high over their foreheads, their wrists heavy with black chronometers. They are tan, and chatter happily about their adventure.

During World War II the U.S. military established a small airbase on Cozumel. The servicemen returned stateside with wild tales of the island's rich, wide reefs. Cozumel, whose Mayan name means "Land of the Swallows," has become, in the offshore Yucatán pantheon, "Land of the Snorkel Bearers," and diving, particularly at world-class Palancar Reef, the major subset of the island's only growth industry, tourism. Available, indeed nearly unavoidable, are dawn dives, morning dives, one-tank dives, two-tank dives, half-day dives, day-long dives, dusk dives, and night dives, all offered by more than a score of dive shops and outfitters.

We opt for the wimp dive, sans scuba, with mask, snorkel, and fins, over a submerged barge that serves—quite well, it seems, judging from the variety and number of fish that use it as their local hangout—as an artificial reef. We splutter and flounder and point wordlessly at what we see: yellow fish; black fish; tawny fish; tan-and-black fish; fish that look like sly buffoons; fish with abrupt, upturned painted mouths that look like Jack Nicholson's Joker; tiny rascal fish; fat fish; thin fish; round fish; stern gray fish; fish that propel themselves as stiffly as Edwardian club members; stately fish, which, betrayed by flapping, nearly vestigial-seeming fins, look like John Cleese demonstrating silly walks; fish blue as electric bug zappers.

COZUMEL, IN 1518, WAS THE SECOND LANDFALL OF THE SPANIARDS WHO PICKED their way west from Cuba and Hispaniola (now the Dominican Republic/Haiti). Isla Mujeres was the first landfall, a year earlier. As the conquistadors forged beachheads and settlements on the peninsula's gulf coast before stabbing their way into Mexico's heart, the island became an afterthought. Except for the odd buccaneer—Jean Laffite and Henry Morgan among them—who used the island as a safe house, Cozumel, little more than a round, flat shank of precocious reef, was left to its own devices.

Truth be told, the island was not that terribly important even to the Mayans. Though it was evidently a shrine of sorts to the fertility goddess Ix-Chel, Cozumel was the Mayan equivalent of a one-horse town, a backwater country church compared to the huge cities and cathedral/temples of Chichén Itzá, Uxmal, and the cliffside peninsula fortress of Tulum, which faces the island.

The history of the Mayans that is available to us is a patchy one, thanks in large part to Diego de Landa, a demonic Bishop of Yucatán who, in inquisitorial fury in 1562, burned nearly every Mayan record he could get his hands on. Though recent discoveries seem to push their beginnings back much further, today's generally accepted evidence puts the first Mayan settlements on the peninsula around 1500 b.c. About 2,400 years later the sophisticated, elaborate, and evidently highly centralized civilization had mysteriously disintegrated into scattered, belligerent parts.

Though it took the Europeans nearly a quarter-century to "pacify" the Yucatán (during the course of the campaign they lost more lives than were consumed during the conquest of the rest of the Mexican mainland), they took to their task with a terrifying vengeance. For 300 years the Mayans were slaves of the Yucatán's plantation economy—cultivating sugar, cotton, tobacco, and maize.

In 1847 the Mayans rose against their oppressors and in a hot and cold war that lasted more than 60 years—during which nearly a third of the peninsula's population were killed—the Mayans fought for, unsuccessfully, their independence. But textbook demarcations are often more precise than reality, and as late as 1924, explorer-scholar Samuel K. Lothrop could assure Yucatán travelers only that "the east coast has gradually become safe for foreigners. . . ."

ONE AFTERNOON AT THE HOTEL, I STRIKE UP A CONVERSATION, IN SPANISH, with an off-duty waiter. Pedro is quite round, and short, with thick curly hair. I ask him if he is an Indian, and he bristles.

"I am a Mayan. Do not say Indian."

I apologize and ask him the nature of my mistake.

"If you call me an Indian, it is like me calling you a gringo. I am a Mayan."

Almost all the hotel's employees, he tells me, are Mayan. This is Mayan land. He speaks forcefully, with obvious bitterness about the *imperio,* the Mexican empire that eventually swallowed the Yucatán, incorporated it.

His mother speaks Mayan exclusively. He speaks Mayan at home, and with his friends and co-workers. But Pedro's children do not speak Mayan at all. There are no schools here that teach Mayan, and no books—at least none that he has ever seen—printed in Mayan. Not the Bible nor the prayer book nor the hymnal. Words such as "sad" and "enemy" and "defeat" spill from him.

But do he and other Mayans get along with Mexicans?

"It is, I think, like the United States with whites and blacks. You are not at war, but you fight. You live in the same country, but you are different people with different lives. And it is difficult to be friends."

Do you speak Spanish with many guests at the hotel?

"No."

Do many guests ask if you are Mayan?

"No. They go to Tulum. They see the ruins. They think we are all dead. Do I look dead to you?"

WE DIVE AGAIN AND SWIM AND EAT AND DIVE. WE ROUND CORNERS ON THE way to and from our room and come face to face with iguanas sunning themselves on ledges. They are large as dachshunds, still as stones.

I SEE PEDRO BEHIND THE BAR SPEAKING WITH A PAIR OF DIVERS. PEDRO AND I exchange greetings and pleasantries in Spanish, and the woman, with one beer in front of her and several more, to judge from her slight slur, inside her, turns to me and thrusts a cassette tape directly in my face.

"You speak Spanish. *You* tell him to play this. It's Jimmy Buffett. We want him to understand our culture. We want him to hear 'Cheeseburger in Paradise.'"

WE DECIDE WE ARE TOO YOUNG TO DIE ON A MOPED WHILE DARTING THROUGH the extemporaneous Cozumel traffic, so instead we engage a taxi for a trip around the island. It is late in the afternoon and our driver is Luís, a thin, mustached hipster with thick, wavy, hair.

Near San Francisco beach, a wide and, of course, white strand fronted by a restaurant-bar, a diving kiosk, and a small curio shop, Luís points to the site of a new cluster of hotels. The largest will have 1,400 rooms, and Luís doesn't think there will be any problem filling each of them.

Summer, he tells us, is the vacation time for Mexican nationals. Winter brings Americans, especially Texans, as well as Canadians, French, Italians, and some Germans.

Who does he like the best?

He is diplomatic. People are the same the world over, and some are good, some are bad. He drives in silence for a while. The Germans are *duros* he says, eventually. Stingy.

WHERE THE HAND OF MAN HAS NOT HAD ITS WAY, COZUMEL, LIKE THE PENINsula's coast, is unremarkable, uninviting. The top soil above the limestone is thin as table linen, and the successful vegetation is tenacious—head-high and ferociously tangled, desperate for moisture.

There were 16 taxis on the island in 1969, according to Luís, and 3,500 permanent residents. Now there are 226 taxis and 70,000 people. There is a depletion of the stores of the sea, Luís says. Fewer conchs, fewer lobsters. He speaks of the days when he could almost reach for a fish from shore. But such is progress. And he would never move to Cancún. It is too hectic there, he assures us.

We swing around the south of the island and break out onto the Caribbean coast, the roily, darkling sea to our right. Luís stops, and we watch some of the locals having a picnic. The men talk and laugh and drink beer, the women prepare food. Children run wild along the sand, their shouts audible above the surf and the rising wind. The gathering, the pleasure, is spontaneous, hearty, very different than the studied, purposeful indolence of the nondiving guests on the rake-groomed hotel grounds of the island.

We detour—Gilberto swept the old road to sea—and stop for a second at an abandoned restaurant, which Gilberto crushed; ditto the "Cup of Coral," a rock formation that Gilberto toppled.

This is a basic road—no guardrails, no road signs, just a cut through the land, a passageway. It is almost dark. The interior is stripped and low. A moor. The mean brush is carved by the prevailing onshore winds. It is gloomy and eerily minimal. But for the first time since our arrival we feel thrilled.

Luís pulls off the road and stops and extinguishes his headlights. Nothing. "There," he says, pointing to the east. "On a clear night, you can see the lights of Cuba."

WE VISIT SAN MIGUEL BRIEFLY. LUÍS HAD TOLD US THAT THE YUCATECAN beers—Leon Negra, Montejo, and Carta Clara—are like no others in the world. "*La leche de una madre*"—mother's milk. The hotels do not serve them. The shops, at least those we tried, did not have them. But there was all the Corona and Tecate and Dos Equis you could drink. Just like back home.

THAT NIGHT AT THE HOTEL, A JEHOVAH'S WITNESS FROM SWEDEN, AFTER HAVing one too many margarita *especiales*, trips on his way back to his own room and falls flat on his face.

Before long the man regains consciousness. He is hardly bleeding. I assist the attentive employees in lifting the man to a wheelchair and rolling him to his room. I ask the assistant assistant manager if this sort of thing happens very often.

"Once a day," is his answer.

On our way back to our room, we stop by the bar and tell people that all is well. A woman wonders out loud if perhaps somebody slipped something in the man's drink.

"Oh yeah," her companion answers. "Tequila."

WE BACKTRACK FROM COZUMEL, BY AIR, TO CANCUN, A 15-MINUTE TRIP, THEN take a cab to Puerto Juárez, just north of town to catch the boat to Isla Mujeres, a six-mile-by-half-mile slip of a thing a dozen miles north of Cancún. The driver is from Mérida, the capital of the state of Yucatán, 175 miles inland. Many people living in these parts, he tells us, have moved from Mérida. For most of the Yucatán's postconquest life, it was the peninsula that mattered economically, first because of food crops, later because of henequen (a type of hemp), chicle, and tropical woods. Since World War II synthetics have made henequen redundant, and overharvesting and the emergence of more exploitable Third World countries have cut into the wood industry. It is the islands, and tourism, that now rule Yucatán's economic roost.

Did you want to leave Mérida? I ask the driver.

"No."

Why did you?

"I had to. There is no work. No work at all."

WE RENT A TINY MOTORCYCLE ON ISLA MUJERES, THE "ISLAND OF WOMEN." (The Spaniards found a lot of female statuary when they landed on the island is one version of the name's origin; or the men were all gone fishing; or the sailors were suffering from having been too long at sea.) We pick our tentative way south, through the boisterous, compact downtown, past a military base where stern young men in blue uniforms, holding rifles, stand guard.

We pass by tortoise pens where, if you have the heart, sad-looking, worn turtles can be observed swimming in captivity. We pass the much-touted Garrafón National Park, which seems to us garish and crowded, and arrive, a few minutes later, at the south point, where we park and walk past an abandoned radar base onto a small headland.

It is windswept here. We poke around a small, chipped, scarred monument—dedicated to what, exactly, is not clear—and investigate a mound of rubble, the remains of some minor Mayan construction that got punched by Gilberto . . . the ruins of a ruin.

Opposite the headland, across the sea is Cancún. The new temples, honeycombed with bedrooms, rise impressively, guard the coast.

We putt-putt our way north along the deserted, windward side, between rolling steppes and the chalky sea. We dawdle and laugh and enjoy ourselves immensely, glad that no person, no guidebook, has deemed this empty quarter worthy of inclusion in their listing of island attractions.

FELIPE IS 19, CHIRPY, WITH HIP, MODESTLY SHAGGED HAIR AND A SMALL CRUCIFIX on a chain around his neck. His first mate on the long, narrow launch we have hired to take us to Isla Contoy (Mayan for "Island of Birds") is Carlos, a quiet, thin, smiling, 28-year-old.

For 45 minutes we bang across the diced sea from Isla Mujeres in Felipe's launch, as fast as custom agents in hot pursuit. Hundreds of cormorants and frigate birds swirl over Contoy's trees as Felipe cuts the engine and baits a pair of lines. Within five minutes they have hauled two dreadful looking three-foot-long barracuda aboard, and pounded them lifeless with a two-by-four.

We tie up at a long, rickety pier stretching out from a lovely, shallow, archetypal bay. We can explore Contoy briefly and return for lunch, Felipe tells us. On the menu is barracuda.

The island, a bird sanctuary with a small museum and an observation tower, is eight miles from land, roughly five miles long, and only a few hundred yards wide. It is a compact, idyllic lesson in insular science. The windward side is straight, rocky, and wave-racked, while the lee is complexly serrated with lagoons and coves, and thick with gray, psychotically gnarled mangrove trees. In their branches are hundreds of immense, sentinel brown pelicans, as ridiculous and stately and likable as penguins or koalas.

The museum's tasteful displays inform us that, in addition to pelicans and frigate birds and cormorants, the island and its waters are home, both temporary and permanent, to spoonbills, red egrets (who look like London punks with spiked pink hair), and 60 or so other bird species, horseshoe crabs, four species of turtles, the omnipresent iguana, and boa constrictors. We step over frantic little chameleonlike lizards. Their bodies are mauve against the sand, brilliant green against low-lying vegetation.

With the exception of the open-air museum's curator and a single, lonely soldier, we are the only people in sight. We eat the delicious barracuda Felipe and Carlos have basted with a piquant red sauce and quietly watch cotton clouds scud across the sky.

Felipe and Carlos say they make this trip most days. Sometimes they detour to another deserted island, nearby Isla Blanca. I say to them that, somehow, from what I had read, this was how I expected Isla Mujeres to be, with the addition of a few hotels and restaurants.

This is how Isla Mujeres once was, they say. "Eight years ago there were 1,500 people on the island. Now there are 15,000. We used to walk everywhere without shoes, without shirts.

"We fished everywhere. It was an island of sand, not pavement. We used to play. Now we work.

"We would wade out from shore and grab lobsters by hand. Then we had to travel around the island, looking for lobsters. Then we began to travel here to find them."

"But you still find them?"

"The lobster fishermen come here with huge nets. They know when the lobsters will be here—they travel on the ocean floor, with the north winds of Christmas. You think there are many birds here? There used to be" Felipe spreads his hands and flaps them, "too many to count. Thousands. Millions. They die in the fishermen's nets."

He grabs his neck and, demonstrating, chokes himself.

OUR HOTEL STANDS ON AN ISLET, ON ISLA MUJERES' NORTH END, AND IS THE fanciest of a score or so on the island. Even so, there is no television in the room, no room service. This is fine, we have been pampered enough.

Guillermo Gros, the hotel's manager, says that the relatively Spartan accommodations are by design.

"People come to Cancún to disappear their work worries, their fast life. People come to Cozumel to play in the water. People come to Isla Mujeres for tranquillity, not television. We don't have a big lobby. People are here to be on the beach. This place is for younger people, informal people. People here are country people, and Cancún is the city. You will hear tales of Cancún—tales of horror—from islanders. I think it is the same as people in your country, from a small town, talking of life in New York City. There is fear of that fast pace."

THE DAY WE LEAVE WE GO FOR A DAWN SWIM. WE SHOP FOR A FEW THINGS. We board the small ferry for the short trip back to the peninsula. A mighty squall hits us before we get three boat lengths beyond the pier, and the captain, a young, athletic fellow wearing a Gotcha T-shirt and luminescent trunks, frantically turns the boat around to tie up against a small tugboat as the pier is obliterated by flannel sheets of rain. His hurried, awkward approach smashes a forward section of the gunwale.

The cabin boy, all of 13 years old, races around trying to close the windows, some of which have glass in them.

The storm subsides. As the captain and a crewman work to repair the damage, two Canadian backpackers abandon ship and walk from the tugboat along a second pier to shore.

A Mayan woman laughs and dries off her granddaughter. Someone mentions that he hadn't noticed any life jackets being issued. Someone else mentions he hadn't noticed any life jackets at all.

LATER, WHEN THE FERRY FINALLY DOCKS AT PUERTO JUAREZ, WE TAKE A CAB to Cancún airport. The cab driver tells us he is from Mexico City and asks us if we want to go to the Mayan fortress of Tulum. We tell him we are heading home. "That's OK," he says. "There's nothing there, really, but a lot of ruined buildings."

HAD WE LOVED MEXICO? PEOPLE ASKED AFTER WE HAD RETURNED HOME. YES. Yes, but we corrected, not Mexico. A Mexico. Tropical, luxurious Mexico. A Mexico of room service and poolside service. Of thick towels and large drinks of colors not found in nature, garnished with fruit carved in the shape of sombreros. Not the Mexico of dusty, rutted roads, of inescapable, crippling poverty. We had been to offshore Yucatán, no more representative, finally, of Mexico, than Aspen, Kailua Kona, or Sanur are of Colorado, Hawaii, or Bali.

We had experienced varying degrees of this luxury, this authenticity. And, if svelte Cancún is not *auténtico*, neither, finally, is frisky Cozumel, nor rural Isla Mujeres. Destination resorts, finally, are not *supposed* to be authentic. They are designed. They provide the comfortable *illusion* of authenticity.

But these darkish thoughts soon gave way to retellings of our last night on Isla Mujeres. We had found ourselves, a little lost, on a neglected side street. We noticed a small menu tacked to a fence post. Behind it were a spacious patio and simple wooden tables. Beyond them, between us and the sea, were low-slung palm trees, lit with colored bulbs, where two small children energetically played tag on the sand.

The only other customers were a merry group of men and women speaking Spanish. The owner-waiter was briefly surprised at our presence but after a moment brought us beer and menus and took our order.

Our food—fillets of fish with a delicate garlic sauce—arrived. Simple. Delicious. We lingered, the only people on earth, fanned by a warm breeze, without a care, in a restaurant whose charms had nothing to do with the present, with any specific decade, any larger world. This, too, may have been some sort of an illusion, but we allowed ourselves to be gloriously, memorably, duped. *Bueno*, we visitors said. *Muy bueno*. [1990]

Cuba: An Elegiac Carnival

PICO IYER

IT IS A MIRACLE OF CALM: Along the whole stretch of empty, brilliant beach, nothing seems to move. There are no hotels in sight, no ice-cream parlors; no radios or holidaymakers or amusement parks. Just the transparent blue-green sea, a few thatched umbrellas, the lapping of the waves. In the distance a single white boat mirrors the blinding whiteness of the sand. Yet even at midday there is an early morning stillness here. Everything looks as silent, as flawless, as a postcard—except over there. Two beefy Soviet workers are wading gingerly though the surf in string vests!

Cuba is, without doubt, the ultimate getaway, especially if it is the United States, or tourists, or the present tense you wish to get away from; though only 90 miles from Florida shores, Cuba seems to have been totally blacked out from our minds. Think of it today, and you most likely think of army fatigues, Marxist slogans, and bearded threats to our peace; you are liable to forget, in short, that Cuba is a tropical island, a Caribbean place of lyricism and light, with music pulsing through the streets, and lemon-yellow, sky-blue, alabaster-white houses shining against a rich blue sea. The long, extended claw that is Fidel Castro's home is, as it happens, the largest island in the Greater Antilles, and, very likely, the least visited. Yet everything that made it America's dream playground 40 years ago is still intact. The water, on a usual day, is 80 degrees; the sun shines an average of more than 11 hours each day. There are 4,500 miles of coastline in Cuba, nearly all of them as tranquil as a private hideaway. Small wonder that Christopher Columbus, laying anchor off this cool-breezed island, pronounced it "the most beautiful land ever seen."

Every time I go to Cuba I come back sounding like a tourist brochure. I bore my friends by counting the ways I love this improbable idyll: a perfect climate (77 degrees on an average day); a many-colored culture vibrant with bohemian dives, troubadour cafés, and a film school partially run by Gabriel García Márquez; and all the warmth of a graceful, passionate, late-night people so openhearted that self-interest and true kindness blur.

Havana days are the softest I have ever seen, the golden light of dusk spangling the cool buildings of the tree-lined streets. Havana nights are the liveliest I know, as dark-eyed, scarlet girls in tight dresses lean against the tail fins of chrome-polished '57 Packards amid the floodlit mango trees of nightclubs like the Tropicana, now in its 51st year of Vegas paganism.

In Cuba the sophistications of Europe dance to the rhythms of Africa, all in a sun-washed Caribbean setting. There is the savor of rum in the bars that Hemingway haunted; the friendly dishevelment of the sea-worn old Mafia hotels, crowded now with dark-featured tourists from Siberia. There is even, in Havana, a Humour Museum.

Yet it is something more than the light-filled surfaces that keeps me coming back to Cuba; for there is in Cuba some indefinable air of adventure and possibility. I never want to sleep in Cuba. And when I return home, I find that it still haunts me like some distant rumba: I can still hear the cigarette-roughened voice of the grandma in Artemisa who took me in from the rain and, over wine in tin cups, spun me strange family tales before leading me across puddles to hear Fidel. I can still taste the strawberry ice cream in Coppelia park, where languorous Lolitas sashay through the night in off-the-shoulder T-shirts. I can still see the round-the-clock turmoil of Carnival, and the Soviet doctor who sat next to me one year, blowing kisses at the dancers. Sometimes, when I go out at night and sit on the seawall alone, feeling the spray of the salt, hearing the faint strumming of acoustic guitars carried on the wind, and seeing the empty boulevards sweeping along the lovely curve of Havana Bay, I feel that I could never know a greater happiness.

In communist Cuba, of course, the visitor finds shortages of everything except ironies. The Bay of Pigs is a beach resort now, and San Juan Hill is famous for its "patio-cabaret." The Isle of Youth, long the most infamous Alcatraz in the Caribbean, now entices visitors with its International Scuba Diving Center. And one beach near Matanzas (the name means "massacres") has, somewhat less than romantically, been christened Playa Yugoslavia. Cuba, in short, has edges and shadows not to be found in most West Indian resorts: The billboards along the beach offer stern admonitions ("The best tan is acquired in movement"), and the gift stores in the hotels sell such light holiday reads as *The CIA in Central America and the Caribbean*. Many things here take on a slightly sinister air. "Cuba's waiting for you," runs the official tourist slogan. "We knew you were coming."

Cubatur's most intriguing attraction may be its daily four-hour excursion to a psychiatric hospital. But when I asked if I could sign up for the tour, the laughing-eyed girl at the desk looked back at me as if I were the madman. "It isn't happening," she said. "Does it ever happen?" I asked. "Never," she said, with a delighted smile.

The real seduction of Cuba, for me, lies precisely in that kind of impromptu makeshift quality, and in the fact that it feels so deserted; the whole island has the ramshackle glamour of an abandoned stage set. Old Havana is a crooked maze of leafy parks and wrought-iron balconies, where men strum guitars in sun-splashed courtyards. Its singular beauty, unmatched throughout the Caribbean, is that it feels as if it has been left behind by history, untouched.

Here, one feels, is all the quaintness of New Orleans with none of the self-admiration. And the freewheeling gaiety of a Sunday afternoon in Lenin Park, where soldiers twirl one another about to the happy rhythms of steel bands, is all the more intoxicating because it is so spontaneous; here, one feels, is all the spendthrift hedonism of Rio with none of the self-consciousness.

Cuba, in fact, is the most infectiously exultant place I have ever seen: It sometimes seems as if the featureless gray blocks of communism have been set down on a sunny, swelling, multicolored quilt so full of life that much of the sauciness of the louche Havana of old keeps peeping through. Let polemicists debate whether the exuberance persists because of the Revolution, or in spite of it: The fact is that the Cubans have made an art form of their appetite for wine, women, and song. One young friend of mine in Havana knows only four words of English, which he repeats like a mantra each day, accompanied each time by a dazzling smile: "Don't Worry! Be Happy!" Very often the island reminds me of that famous statement of the 18th-century Englishman Oliver Edwards: "I have tried too in my time to be a philosopher; but I don't know how, cheerfulness was always breaking in."

This exhilarating sense of openness hit me the minute I landed in Havana on my most recent trip: The customs officials in the airport were dressed in khaki but winkingly turned the other eye whenever they saw cases piled high with 15 pairs of new, ready-for-the-black-market jeans;

the immigration officials, when not cross-examining tourists, made kissing noises at their female colleagues. Out in the streets I was instantly back inside some romantic thriller, with intimations of crimes and liaisons in the air. Dolled-up señoritas looked at me with the sly intimacy of long-lost friends; rum-husky men invited me into their lives.

By the following night, I was sitting along the seawall with a group of earnest young students eager to thrash out Hermann Hesse, Tracy Chapman, yoga, Henry Fielding, and liberation theology. Later, walking through the commercial buildings of La Rampa, I heard the joyous rasp of a saxophone and, following my ears through the video banks and rainbowed portraits of the Cuba pavilion, found myself standing in a huge open-air disco, free (like most museums, concerts, and ball games in Cuba), and alive with teenagers in "We stick to Fidel" headbands and Che Guevara T-shirts dancing to a Springsteenish band. In this way—the government apparently hopes—are party-loving kids turned into Party-loving comrades.

When the concert ended, about midnight, I walked over to the ten-stool bar in the Hotel Nacional, where four cheery, red-faced Soviets were singing melancholy Russian ballads to a flirty *mulatta* of quick charm. The girl counted off a few figures on her long pink nails, then swiveled into action.

"Ivan, Ivan," she cooed across at a lugubrious-looking reveler, "why don't you dance with me? Ivan, don't you like me?" At which Ivan lumbered up, popped a coin into the prehistoric Wurlitzer, and, as "Guantanamera" came up, threw his hands in the air and began wriggling in place with all the unlikely grace of a bear in a John Travolta suit. This, I realized, was not Club Med.

The country's beaches—289 of them in all—begin just 20 minutes from the capital. At Santa Maria del Mar, a virtual suburb of Havana, lies one of the loveliest, and emptiest, strips of sand you'll ever see, with only a few old men—salty castaways from Hemingway—standing bare chested in the water, trousers rolled up to their knees, reeling in silver fish. Behind them, across a road, reclines a typical Cuban seaside hotel, filled as always with something of the plaintiveness of an Olympics facility two decades after the games have ended. Next to once-futuristic ramps, bulletin boards crowded with happily crayoned notices invite foreigners to "Workers Shows" ("a very nice activity," offers one board, "where you will see the workers become artists for your pleasure"). Every Monday at 4:30 there are "Cocktail Lessons," and every afternoon "Music, Dance and Many Surprises." But when I looked at my watch, I realized that it was 4:45, and Monday, and not a cocktail student, not a sign of music or dance, was to be seen; somehow Cuba is always out of season.

The proudest attraction of the tourist office—and its brightest hope for gaining needed dollars—is the string of coral keys that sparkle like teardrops off the coast. One day I took the daily flight to Cayo Largo, an absurdly beautiful stretch of 15 miles of open beach, graced with every enticement this side of Lauren Bacall. As soon as I got off the plane, at 8:45 A.M., I was greeted with a frenzied Cuban dance band and a lobster cocktail; for the rest of the day, I simply lay on the beach and gasped at the cloudless line of tropical colors—aqua and emerald and milky green, flawless as a Bacardi ad. There is nothing much to see in Cayo Largo, save for some basins full of turtles and an islet featuring 250 iguanas; but the government is now organizing deep-sea fishing, nighttime scuba diving, and marlin fishing to bait the tourists. And as with all the most delectable resorts in Cuba, the place is utterly uncluttered, in large part because locals are not permitted on the beach. (This is no legal fiction: I myself, while walking along the beach one drowsy Sunday morning, was hauled over by a policeman hiding in the bushes, on grounds of impersonating a Cuban.)

Recently, in a bid to rescue its shattered economy, Cuba has begun refurbishing its old hotels with tiled patios and stained glass, and enticing visitors from the travel department of *Vogue*; but even now, thank Marx, the island remains roughly 90 percent tourist-proof. One still needs two chits and a passport to buy a Coca-Cola and, as in some loony lottery, Visa cards are accepted only if they contain certain numbers. This, though, is part of the delight of the place: Whenever one goes out—day or night—one never knows how the excursion will end, or when.

One sleepy Sunday not so long ago, I waited for a taxi to take me back to Havana from the beach. And waited, and waited, and waited, for more than three hours in all, under a tree, on a hot afternoon. Finally, just as I was about to lose all hope, up lurched a coughing, red-and-white 1952 Plymouth, with "The Vampire Road" written across the windshield. Seven of us piled into the wreck, and the next thing I knew, the quartet in back was pounding out an ad hoc beat on the seat and breaking into an a cappella melody of their own invention—"*Ba ba ba, we're going to Havana . . . ba, ba, ba, in a really sick old car.*" For the next two hours the increasingly out-of-tune singers unsteadily passed a huge bottle of rum back and forth and shouted out songs of an indeterminate obscenity while the walrus-mustached driver poked me in the ribs and cackled with delight.

In Santiago de Cuba, the second city of the island and the cradle of the Revolution, I spent a few days in the gutted home of a former captain of Fidel's. From the hills above, where Castro and his guerrillas once gathered, the city looked as it might on some ancient, yellowing Spanish map. Down below, though, in a peeling room that I shared with a snuffling wild pig that was due to be my dinner, things were decidedly less exalted. Every night, in the half-lit gloom of his bare, high-ceilinged room, decorated only with a few black-and-white snapshots of his youth, my host took me aside ("Let me tell you, Pico Eagle . . . ") and told me stories of the Revolution, then delivered heartbroken obituaries for his country. Next door, in an even darker room, his son prepared dolls for a *santería* ceremony, the local equivalent of voodoo. And when it came time for me to leave, the old man asked me for some baseball magazines from the States. Any special kind, I asked? "No," he said softly. "I like the ones with Jackie Robinson in them."

That sense of wistfulness, of a life arrested in midbreath, is everywhere in Cuba: in the brochures of the once elegant Riviera Hotel that now, disconcertingly, offer a "diaphanous dining-room"; in the boarded-up stores whose names conjure up a vanished era of cosmopolitanism—the Sublime, the Fin-de-Siecle, Roseland Indochina; in the Esperanto Association that stands across from a dingy, closed-off building under the forlorn legend "R.C.A. Victor." Hemingway's house in the hills is kept exactly the way he left it at his death almost 30 years ago—unread copies of *Field* and *Sports Illustrated* scattered across his bed. And the buildings all around, unpainted, unrepaired, speak also of departed hopes. One reason why so many Cubans ask a foreigner, *¿Qué hora es?* is to strike up a conversation—and a deal. Another, though, is that they really do need to know the time in a place where all the clocks are stopped.

Perhaps the most haunting site in the beach resort of Varadero is Las Américas, the lonely mansion above the sea built by the Du Ponts. Nowadays it is a dilapidated boarding school of a place, filled with long corridors and locked doors. The Carrara marble floor is thick with dust, and the photos in the drawing room are hard to make out in the feeble light. But along the mahogany and cedar walls there still hangs a tapestry poignantly transcribing all the lines of the poem that contains the hopes the home once embodied:

"In Xanadu did Kubla Khan
A stately pleasure dome decree . . . "

It is that mix of elegy and Carnival that defines Cuba for me, and it is that sense of sunlit sadness that makes it, in the end, the most emotionally involving—and unsettling—place I know; Cuba catches my heart, and then makes me count the cost of that enchantment. Cuba is old ladies in rocking chairs, on their verandas in the twilight, dabbing their eyes as their grandchildren explain their latest dreams of escape.

It is pretty, laughing kids dancing all night in the boisterous cabarets and then confiding, matter-of-factly, "Our lives here are like in Dante's *Inferno.*" It is smiles, and open doors, and policemen lurking in the corners; and lazy days on ill-paved streets; and a friend who asks if he might possibly steal my passport.

In Cuba the tourist's exciting adventures have stakes he cannot fathom. And every encounter only leaves one deeper in the shadows. My first night in a big hotel, a girl I had never met rang me up and asked, sight unseen, if I would marry her. The next day, in the cathedral, a small old man with shining eyes came up to me and began talking of his family, his faith, his grade-school daughter. "I call her Elizabeth," he said, "like a queen." He paused. "A poor queen"—he smiled ruefully—"but to me she is still a queen." When we met again, at an Easter Sunday mass, he gave me Mother's Day gifts for my mother and, moist-eyed, a letter for his own mother in the States. It was only much later that I found the letter was in fact addressed to the State Department, and the kindly old man a would-be defector.

And one sunny afternoon in a dark Havana bar, so dark that I could not see my companion's face except when she lit a match for her cigarette, I asked a friend if I could send her anything from the States. Not really, she said, this intelligent 23-year-old who knew me well. Just a Donald Duck sticker for her fridge. Nothing else, I asked? Well, maybe a Mickey Mouse postcard. That was quite a status symbol. And that was all? Yes, she said—oh, and one more thing: a job, please, with the CIA.
[1990]

Martinique: Island of the Returning Ghosts

HERBERT GOLD

THE PEACEFUL ARAWAK INdians lived here first. Then the Caribs paddled up in their little boats and ate some of the Arawaks, married others, and enslaved the rest. And then the Spaniards and the French did the same to the Caribs (who now had a little Arawak blood running in them), only the Europeans acted without much of the cannibalism, since they were Christian and this was opposed to their moral law and dietary custom. Still, they did enslave and did intermarry.

Necessity required that the Spanish and the French bring new slaves from West Africa to grow sugar, a labor that the Caribs refused to the point of briskly dying under the whip and gun. These former Africans still do the work of Martinique and live the pleasures of Martinique; and, although they're called *Martiniquais*, they have a little Carib and Arawak in them, and quite a lot of the French planters in them, plus some wandering Dutch, Dane, German, Chinese, East Indian—and pirate, trader, and second sons of miscellaneous ancestry.

Martinique is an overseas department—not a colony—of France. "Negritude," the term made popular by Frantz Fanon and Sartre, was taken from the poetry of Aimé Cesaire—surrealist, former Communist, once mayor of Fort-de-France—who used it to describe the black awareness of their isolation and oppression. It is a symbol of revulsion against whiteness, against colonial oppression, against European ways of working at life. The Martiniquais seem to want the goodies of this world, perhaps the goodies of heaven, more than they want world revolution or even independence in Martinique. Socially they are afflicted, as the underclass is everywhere; but legally they have equality with every Frenchman, and opportunities at least halfway open. Visiting radicals would like to find cells of revolt in Martinique. Instead they find complaints and kindness and an unmurderous people doing the mild best it can.

WHAT IS A FUR HAT DOING ON THE HEAD OF THAT CRUMBLING GIANT BEARING a flag that represents France, Africa, and Martinique, an armful of flowers and a sheaf of leaflets headed THE DOGMA OF HAM?

Why don't I just ask him?

"The fur hat, my dear visitor from overseas, represents the reconciliation of Catholicism with Christianity. I am the martyr interned by the traitors of Vichy. Now I am the founder of the Dogma of Ham. Excuse me, esteemed sir, I must present the flag."

We are in Fort-de-France at a ceremony one Sunday morning in June that is devoted to the anniversary of the Call to Resist of General de Gaulle. After a ceremonial mass in the cathedral, the heroes of several campaigns—the halt, the lame, the blind, the old—will gather before the monument dedicated to "the Children of Martinique Fallen for France." While we wait, the paras, the florists, the police, the children, the priest, the heroes make ready. An engineer, fingering his medals, tells me he has passed the examination for promotion, but "they" won't promote him. Together we watch the honor guard form—a pipe-smoking, white-gloved French officer in his kepi; a tough, blond para noncom, jaw twitching, in beret and camouflage suit; a company of black and white paras; and, opposite, the small group of aging, bemedaled black heroes who are about to be honored. Medals eaten by years of tropical salt humidity glint in the sunlight. One hero is elegantly leaning against his cane, tapping his dress gloves—vinyl American driving gloves.

Finally a Peugeot 604 drives up with the French prefect in his blazing whites, the two groups of soldiers salute, the trumpet sounds, the motor of the Peugeot 604 is running, the prefect salutes everywhere, a black hero with flashing medals reads the Appeal of General de Gaulle, there is another blast of trumpets, the fur-hatted Dogma-of-Ham man presents the flag, a trumpet plays taps, the motor of the Peugeot 604 is still running, a gray-bearded old priest behind a mosaic breastplate of medals invites me to visit his cabinet, the elegant, non-sweating prefect pauses an instant at the door of his Peugeot 604 and it opens, the prefect departs and the ceremony is over.

The Panasonic transistor radio next to me is playing a Creole song,

Oh Cherie,
Three times a day!

about the love of somebody for somebody else. On the beach a lovely, laughing teenager toys carelessly with the affection and metabolism of a young lad. They are somebody and somebody else.

The heroes of several wars, in their black suits and canes and gloves and medals, reluctantly disperse.

It is a fine and peaceful tropical morning in Fort-de-France, Martinique, "Queen of the Antilles." The ceremony takes about ten minutes, once a year.

IN THE DEPARTMENTAL MUSEUM NEAR THE SAVANE IN FORT-DE-FRANCE, NOT far from the statue of the beautiful Creole, Josephine, Empress of France, I visited the room dedicated to:

The African slaves
our ancestors
first artisans
in chains
of the prosperity
of this island

and stood before the rusted slave collar attached to its heavy chain. In this room there are also indulgent, tinted portraits of whippings, of recalcitrant Africans being heaved overboard—and of rich plantations and sugar refineries—and, behind the glass, the gold skirt of a slave girl emancipated by her lover.

Upstairs I visited a traditional Salon Martiniquais while outside I heard the buses (called "les cars") which are painted in cream and green, miniature Paris buses, little buses for *les petits Français*.

Then I walked up the hill, past walls spray-painted with the words FRENCH OUT OF AFRICA! EUROPE OUT OF AFRICA! to visit an old friend, Edouard Glissant, novelist, poet, critic, playwright, editor, scholar, teacher, philosopher, ethnologist. He has advanced degrees, he lectures and is published worldwide, his wife is a doctor, they live well, they have three handsome children. Both Edouard and his wife come from rural stock and are the first of their families to enjoy a higher education.

A Haitian maid serves us lunch at home; after, we look at the press clippings for Glissant's play, "Monsieur Toussaint," a success in Paris. Solicitously and fondly he inquires of each of his children how the morning at school has gone; later, in his Citroen, we will drive them back.

It would seem that the Glissant family is a model for what France can do for her sons and daughters. But Edouard Glissant does not see it this way. "The English want black people to be separate and unlike them—that's their prejudice speaking. The French prejudice is to make us be like them."

Glissant is a nationalist. He is not among those who pretend to be French, or those like Aimé Cesaire, who now seek autonomy within the French system. He does not like to hear about the blasted states of Africa with their tribal wars and manipulation by money and power from elsewhere. He smiles ironically about the legal fantasy which makes Martinique, thousands of miles from Europe, a department of France. "Department," to him, is "colony."

"There is forty percent unemployment," he says. "The French system of official lying and the laws of social assistance hide this. The misery is in the soul of people who do not manage their own lives."

When I ask if independence would end unemployment, he replies that Martinique is supported by France because of its resources. The sea is rich. The French are not ones for self-sacrificing generosity. But even if another island nation of half a million were to join so many other confused members of the United Nations, he says, "Look, my neighbors are békés." (This is the word for the old white French of Martinique, ten or so families, some of them rich planters, much intermarried. There are also the "guava békés," the poor and the fallen among this traditional and proud elite.) "They can't understand how I, a black writer, married to a black doctor, can live among them. People who send their children to the school I direct would not let them marry anyone the color of my children. Some of them will see no doctor but my wife, and yet they are békés and we are black. Of course, the békés don't like the French, either. The cane sugar people don't like the beet sugar people. *Bon jour*!"

"Would it make trouble for you if I write this?"

"They know what I think. Write away, *mon vieux*."

MARTINIQUAIS SOMETIMES FEEL COLONIZED, THEY FEEL AS IF THEIRS IS THE NAtion of Négritude and not France, they do not feel like "little Frenchmen." Martinique, a department of France, is separated from, say, Brittany, by one hell of a lot of water, land and history.

Cruising the Caribbean, Christopher Columbus wrote, "It is the best, the richest, the sweetest, the evenest, the most charming country in the world. It is the finest thing that I have ever seen, and therefore I cannot tire myself of looking at such greenness." France enjoyed the islands, too, and in return the Empress Josephine enjoyed France. But the pleasure has not always been mutual. Mount Pelée, source of the greatest volcanic disaster since Pompeii: an eruption in 1902 that killed over

30,000 Martiniquais, still broods over a people determined to make every day a festival, every day to be lived as if it might be the last. Tropical paradises nowadays seem to be islands which *decide* to make themselves paradises, and there's a taint of caffeine nervousness in the intention. The brooding, befogged immense peak—over 4,000 feet high—stands like a wicked god over this island, dominating it geographically and perhaps psychologically. Négritude and Volcanic Fever have entered the personal equation along with the usual peculiar relationship of subject peoples to their French guides.

And yet, surprisingly, little sense of rebellion, riot or tension appears in daily life. The country is densely populated, producing sugarcane, bananas, pineapples, and tourism for export; other tropical fruits and vegetables for local use. It sends deputies and senators to the French Parliament. It receives the benefits of the somewhat eroded Napoleonic Code. It speaks French in school and elsewhere, Creole, that savory mixture of French with African languages, and an occasional vagrant Spanish or English word, thanks to pirates or traders. It sort of runs itself, under a prefect appointed by the minister of the interior in France, but of course that "sort of" covers a bit of fudging. Martinique scampers behind Paris on a long leash, but on a leash nonetheless. "True democracy would consist in not raising some Martiniquais above others," a French official remarked, grumbling about the number of educated and professional people of African aspect in this "overseas department."

Despite its burgeoning tourism, Martinique belongs to an older world; it still looks like the slave coasts of west Africa, with swaying *marchandes,* intense marketing and storytelling, a world open on the surface, colorful and noisy, and closed beneath. Symbolizing this equilibrium, most Martiniquais speak French with visitors, and speak the good French imparted by general education. But when they turn to each other—to make love, to argue politics, to comment on the stranger, to joke, for the important transactions of living—they adopt the Creole which is impenetrable to most visitors. When I sometimes replied in Creole (I lived in Haiti for some years), their faces ignited with the comedy of it all. "White man there speaks Creole! Hey, look! And thinks he's smart!" And gusts of laughter about the peculiar hubris of *zoreilles*—a Martiniquais word for the French stranger who, in this case, happens to be a North American. The word sounds to me (amateur philologist) like the Creole equivalent of the American Chinese "round eyes" for Caucasians.

Columbus landed at Le Carbet in 1502, needing four trips to the Americas to find Martinique. Near here, in Turin Cove, Gauguin lived and painted for four months in 1887. Martinique seems *nearly* to be a part of history, as she is *nearly* a part of France.

Her people are of the real world. From the cafés near the Savane on the waterfront of Fort-de-France, or on the back streets of Paris which seem to have made a tropical dream transport into the crowded, winding, balconied ruelles downtown, the capital city has a worldly French, African, and Caribbean bustle. The hills beyond, shrouded in mists, are mysterious and lovely.

Like all cities, Fort-de-France is a place of exchange—goods, money, services, bodies. Aristotle said we come to cities in order to live and also to live well. The people of Fort-de-France are followers of Aristotle without necessarily reading him. They may have neglected to appoint virtuous men to govern—Plato's rule for a healthy city—they have let others appoint those who govern them. They don't read Plato, either. They read *Paris Match*. Some read *France-Antilles* for news of soccer and sugar prices. But they do live pretty well. The fruit of the sea and the fruit of the plain and orchard and the grape of the vine and the rum of the sugar—especially the rum of the sugar—and living in the shadow of the catastrophes of nature. Occasionally there are hurricanes. Once there was a holocaust from Mount Pelée to remind the living of the precariousness and transitoriness of life. For the people of Saint Pierre, the reminder of 1902 was instant and final. Six thousand people now live amid the volcanic melt; 30,000 died, every single soul of Saint Pierre except for a convict buried underground in a dungeon. Saint Pierre, where contorted clocks still show the time of the explosion, and wine bottles took shapes glassblowers have never conceived, is no longer known as the Little Paris of the New World. The volcano is carefully monitored by experts.

Romantic travelers have called Martinique *le pays des Revenants*. The white families that controlled the island, French and traditional and rich, have intermarried ferociously, and some even report they now have "the blood of the country"—African, East Indian, ghostly Carib—running in their veins.

Whatever metaphysical speculations about bloodlines, simple observation confirms something rare enough and practical: rum, coffee and pleasure run in the veins of a people calibrated to the end games of sun, volcano, island limits, and onrushing time. Living under the regime of an "overseas department of France," with the energies of politics a compelling option only for a few intellectuals like Edouard Glissant, dancing, talk, and sexuality occupy the spirit rather powerfully. West Indian languor is a reality, modulated by the stimulants of pleasure. The skepticism of the French, plus the mortal oppression of slavery and poverty, plus the ominous history of nature in this beautiful place, have given the Martiniquais a set of hairline emotions. They accept, they smile, they dance, they play, they represent a seductive island in human nature. And there is melancholy and intelligence in the languor. One cynical young man, who studied American history at a German university, said: "We have government of sugar, for sugar, and by sugar."

As in Haiti and in old New Orleans, a complex and pedantic vocabulary has been developed to describe the modulated varieties of color and blood of which "the three races" of Martinique—black, white, and mulatto—are very crude summaries. Such terms as *capresse, griffe, quarteronne, metisse, chabine* express finicky distinctions having to do with color, springiness of hair, broadness of nose, prominence of lip, etcetera. Even the etceteras are discussed in detail by snobbish connoisseurs. It is all additionally complicated by the fact that a dark millionaire is seen as fair, a fair porter is seen as dark. The optic that measures East Indian, African and European bloodlines is as important as the bloodline itself. In some ways, in a place dominated by people of color, this seems a harmless hobby. To the visitor, the elegance of the peasant woman on the road or the laborer in the field suggests that the combinations are good, and the melancholy eyes and straight black hair that one also finds make it seem likely that some ghosts of the extinguished Arawaks and Caribs have found their way home.

The casualness and insouciance of the people of Martinique, Edouard Glissant has written, come from a deep sense that they are just here in passing, no longer belonging to Africa, not ever a part of France, just floating on their volcano through time and space, with no direction home. (On the radio I heard a group of African voices, accompanied by ritual Congo drums, singing a Creole version of "Mr. Tambourine Man," a popular song by the Jewish American poet/composer, Bob Dylan.)

Despite the surface peacefulness of life in Martinique—reminiscent of Tahiti, another flawed French jewel—there are crowded slums in Fort-de-France and there are underemployed country workers who wish for more than rum and dreaming and making little ones. During Vichy days, the French naval rule of the island was brutal and openly prejudiced—food taken from black given to white—and some observers credit the persistent Communist party and occasional eruptions of rioting or anti-white feeling to a certain insecurity: this could happen again. But for the moment, there is a precarious exchange: the rich landowners exploit the poor, and the poor use the paternalistic delicacy of France. It is not a balanced deal, but it seems to work just now. A French colonel, met at a picnic on the beach near Diamond Head, told me that the youth of Martinique are neater, cleaner, politer, and smilier than Paris youth, and he prefers to rule them.

The extended family, an African tradition, with love given those who care to receive it, makes this a non-hate-filled world; indeed, a tender world. Without sentimentality, one can report good vibes on every road, in every village, at every hour. Grand Rivière and Basse Pointe in the North are immaculate; the people of Ajoupa Bouillon and Morne Rouge in the mountainous interior smile, wave and joke with the dusty traveler; the cemeteries are filled with flowers carried by pretty girls, and it sometimes seems that the poor make their precarious living selling each other flowers.

Injustice exists. The response of capital and bureaucracy is inadequate. But hate does not run in the streets.

I tried to talk with a journalist about politics and anger. Instead, he described a girl I had met as "cold" because she has only a few lovers. There was much greater concern with sexuality than with revolution. He suggested that I not bother discussing politics with the lady. He may have been reflecting the political fact that, traditionally, any béké or white foreign male can seduce a black woman, despite her instinct for virtue. The economic and social advantages of a white lover and the consequent mulatto children are too great for poor black girls to resist. ("I'd like you to meet my son, the mulatto.") As in Haiti, there are several types of menage—from Christian marriage sanctified by the church, to common-law marriage, to *placage*—a kind of recognized, statused polygamy—to concubinage, to simple catting around. The menage has many forms, and desire leads us into a maze. We may be preoccupied with it, but our chief expression of the preoccupation is with laughter. Children give pleasure. We can hope the woman will continue to give pleasure. And if she doesn't, another surely will.

THERE WERE TWO SERIOUS SLAVE REVOLTS IN MARTINIQUE AFTER THE SUCCESSful Haitian rebellion of 1804. The viability of the slave economy declined, until finally the slaves were freed in line with the general post-Napoleonic national liberations of Europe. France was tired of paying an army of 3,000 to control 70,000 slaves for the benefit of 9,000 whites. East Indians, Chinese and Annamites came to work as indentured servants, along with the Africans known as *les Congos*. In the last hundred years all the races have been combined to produce a population of nearly one-half million people.

In very early days, a white father could free his mulatto child by paying a ransom to the church. The shortage of white women encouraged cultivation of black ones. Later, slaves were given certain privileges by the church—they could be buried in a holy cemetery, instead of a field, if they were baptized. Death, flogging and mutilation were normal punishments for slaves who misbehaved. The slave was a *thing*: perceived as imperfectly human. Popular belief held that the *blanc* is the child of God, the *noir* the child of the devil, and the mulatto has no father. In the eighteenth century even certain clothes were forbidden the person of color. All the marvelous resources of the civilized mind—distinction, definition, ranking and ordering—were bent to this fearful and manipulative and oppressive spite. At all costs, the white man must reign.

Things are different now. People of color occupy political posts and fill professional jobs and even get rich. But the spirit of the past lingers on, with the mulatto tending to replace the white, and the black still at the bottom of the social heap. Finally, compulsory education and the pressure of history are making Martinique a place where only normal xenophobia interferes with progress. France still pulls the strings.

Land division follows the pattern left by slave times. The major planters have large plantations. The very small holdings are near the villages of Rivière Salée, St. Esprit, Vauclin, Morne-des-Esses, Lorraine, where escaped slaves and freed mulatto children found refuge. The fertile tracts are held by the major families. The poor land is held by peasants, often descendants of families who, in one convulsion, gained freedom and a spot of volcanic soil. Today, land tends to be subdivided into economic uselessness.

So the movement here too, in this pre-industrial society, is to the city. An urban proletariat, with nothing to be proletariat about, has moved away from its traditional small farming. In the country, a black man works the fields and lives with his family. In the city, if he serves the tourist industry, say, he may begin to see himself as a mulatto. If he prospers, others too will see him as a mulatto.

He will never be seen as a béké—the ruling families who have been there since colonial times and never married a black in church. They marry each other and try to keep from degenerating. They export sugar, rum, pineapples and bananas. Many of them are very rich, and a béké, even a poor one, remains a member of this special, miniscule, psychologically privileged class. A rich mulatto is perceived as fair, but it is impossible for him to marry a béké; if he does marry white, his children are still mulattos, not békés. Oh, complication. As in Haiti, the color line is drawn in unfamiliar ways but it is drawn with the utmost care.

The levels of society, then, include the békés at the top, ostensibly pure white, with the values of a traditional feudal aristocracy, inheriting land and religion; the prosperous mulattos, who fill many of the professional and governmental jobs; and then the seventy percent of the population that is black, poor, sugarcane workers or urban proletariat. A few mulattos belong to this class by affiliation and a very few outcast whites. At the very bottom are the *gens cases*, a blend of sharecropper and serf, who live on plantations and do not wander in search of other work or a plot of their own. Poor fishermen also belong to these bottom dwellers.

While some blacks climb into professional or business success, they are still not seen as "elite." Within each color stratification there are gradations according to wealth, power, education; although this stratification is fraying at the edges where nature has taken its course and the colors blend. Because of this, békés and mulattos meet in business, not socially; and mulattos, with less intensity, limit their social contacts with blacks. Béké men, but not their women, associate socially with mulattos. When intermingling between white and color occurs, the whites are likely to be immigrants from metropolitan France.

AS A FORMER RESIDENT OF HAITI, A FREQUENT VISITOR TO THAT LOVELY, haunted place, I wanted to compare the effect on similar populations, geographical situations and poverty of two forms of government: the anarchic corruption and spasmodic tyranny of Haiti, the colonial management of Martinique. The conclusion is not so clear as it might be. Haiti, with all its painful history—in which, typically, a "provisional president" could order his political opponents "provisionally shot"—has the pride in nationality and a vigorous culture that independence can give a people. The slightly barbered look of the exotic in Martinique bespeaks both the economic aid and the heavy hand of France. If I wanted to live well, I would surely live in Martinique, where there is beauty, grace, comfort, and entertainment. But if I wanted a dramatic life, one of risk and passion, I might prefer the dangerous currents of voodoo Haiti.

In Martinique, there are battles of vipers and mongeese, cockfights, violence for show, in the Latin fashion. But church and state have made sure that the Martiniquais, with their madras exoticism and compulsory French education, remain *des petit Francais*—which is not the same as to be *French.* Not so long ago the church assured its slave clients that serving a master was good practice for serving God, and that heaven would be soon enough for justice.

The combination of Africa and France certainly works for good eating. West Indian and Caribbean cooking has a touch of whimsical invention: octopus, iguana, bat, for the adventurous; crispy baguettes for those nostalgic for Paris mornings; local fruits for the average lover of fresh, sun-ripened bananas, papaya, guava, oranges. An American I met, raised in Los Angeles on United Fruit bananas shipped green and ripened with the aid of mustard gas spray, asked fretfully: "Don't these bananas taste funny?"

"They won't hurt you," I assured him. "It's only sunlight."

He looked worried. There is that peculiar clarity in the air, mixed with a tang of charcoal cooking. "The lack of smog won't hurt you. You don't need particles and photochemicals to protect your skin from God's sunlight."

As opposed to Haiti, the blacks of Martinique read and write, they find a relatively complex technology both in the cities and on the plantations, they move easily between town and land, their Christianity is firm and the practices of magic, voodoo and superstition are limited. Echoes of Africa persist in extended family structure and the powerful loyalty to local communities. Ironically, surviving folk tales from Africa, dances, tastes in spicy food and music are seen as consumer products. A black lawyer, playing a goatskin drum at a family party, finishes, sweating and laughing, and asks the American visitor: "Would you say I have a natural sense of rhythm?" And when I hesitate to reply, he says: "*Res ipsa loquitur.*" And laughs a thigh-slapping Sunday laugh, not the discreet Gallic legal snicker of his Monday-to-Friday life.

In a village, the house may have chickens wandering in and out. There is a family pig. Madame sells rum: She is very African, this commercial wife—she may also sell thread, fruit or an old sink at her front door. She also serves her man his meals, and the children eat standing up, at no fixed time, when the food is ready. It may be a straw house, patched with mud and newspapers, decorated with religious artifacts or photographs; or it may be a cement house with a bamboo porch built with the aid of friends from materials collected in the forest.

Most likely, these people bathe every day in a stream. They carry their water from a village well. Perhaps they have electricity; perhaps not. If they do, the students can study by the bare electric bulb. If not, they do the best they can—under streetlights in a village. Oral storytelling survives in the antiphonal chanting and answering of *Cric! Crac!* "It happened one time in a land far away . . . " The beat of the drums seems to be for dancing and play, not ceremonial for religious purposes, but the matched battery of drums is similar to African and Haitian drums.

In Haiti there are *houngans*, voodoo priests. In Martinique, the closest to the houngan is the *quimboiseur*, the magic man. He sells charms, healing remedies, perhaps aphrodisiacs. He explains sudden deaths or turns of evil luck. He may make women fall in love; he may punish a man for abandoning a lady. In Haiti, the *ouanga*—a doll to be tormented so that the person it represents will suffer. In Martinique, the *bois-bois*—an effigy that can be burnt or have pins stuck in it. These things persist although now illegal. But, as the folk saying puts it, "In Martinique, we do not like to speak the truth when it is bad."

One morning I stopped at a hut by the road at some nameless place in the north. Inside, a man was playing dominoes by himself, slapping the tokens hard, conversing intently with the white and black spots. Yes, he lived alone. No, he did not entirely like it that way. But he had asked the stars and he was not supposed to marry. "It's not in my planets," he said.

"Well, move the planets," I suggested.

That seemed like an interesting idea. He considered it.

"If I go to the quimboiseur, he will make my crop prosper." He gazed mournfully into my eyes at the injustice of this. Then he smiled and touched me, tapping my belly like a rural Frenchman. "However, I had better plant it in fertile soil first."

Haiti was called the Pearl of the Antilles. Martinique calls itself the Queen of the Antilles. They have more than this small braggadocio in common, despite the significant difference of chaotic self-government and elastic but firm French paternalism. Both are mountainous and small—Martinique about forty miles long, half of that in average width—and both are volcanic in origin, with terrain mightily crumpled. In Haiti it is said that farmers sometimes die falling out of their cornfields. The terraced spaces of Martinique are almost as steep.

The market women trudge through the country with burdens on their heads. Even on "les cars," they seem to be loaded with chickens, goats, cloth or tin handicrafts. As in Haiti, the men carry less or nothing at all.

SO THE CARIB INDIANS REPLACED THE ARAWAKS. THE SPANISH DROPPED BY TO hunt treasure in 1552, and promptly left again. The French came in 1635 and stayed, although the British tried an invasion or two, were chased off, leaving few English words embalmed in the local Creole. First the French brought religious fervor and tobacco farmers from Dominica; later they cultivated slavery and sugar. Since the French tradition for slavery was not so highly developed as the British, they also brought indentured servants, and these Caucasians either disappeared or intermarried. The Dutch advised on sugar production and helped promote more efficient slaving. By the middle of the eighteenth century, the population reached approximately 16,000 whites, with perhaps three times that many blacks, but by the time of the French revolution there were only 10,000, with at least eight times as many slaves. Emancipation sped up the process of blending these colors and conditions.

Originally, the slaves were brought to the Antilles to work the plantations. The combination of fertile land and captive labor produced great riches and abominable social conditions. Martinique still imports most of its essential commodities; it is still a plantation society. A French study admits that seventy percent of Martinique's people live no better now than they did under slavery. It is not sweet to work the cane fields.

Yet good communications, a close tie with France and the prevalence of the church have meant that Martinique seems almost, with its near, swept villages and farms, like a tropical province of the Metropole. The 385 square miles of fertile volcanic earth are not so desperately poor as Haiti, where the standard of living is now lower than in slave times. If Haiti is an anarchy, Martinique has achieved the stage of feudal society: kings, masters, priests, serfs, with assigned functions; the bare beginnings of a local middle class, a commercial, artist and professional class.

A curious byway to explore is the persistent Jewish presence in the Antilles. In 1654, Benjamin da Costa d'Andrade arrived in Martinique from Brazil, and by 1676 there was a synagogue, with a Torah brought from Amsterdam by d'Andrade. Later, the vicissitudes of persecution and changing political fortunes obliterated the formal Jewish community, but some light-skinned Martiniquais still bear Jewish names and aspects. Marrano Jews found their way here from Spain via Holland and South America.

Now a few of the tradesmen and professionals from France are Jews, but many more of them are *pieds noirs* from North Africa who have never lived in the Metropole and have the habit of adapting their lives to the margins of a native society. Near the waterfront of Fort-de-France, a blast of air conditioning from the doorways of the *bar snacks* or the *mode de Paris* shops welcomes the stroller to a Mediterranean style, and the olive-skinned, olive-eyed children in shorts, clinging to mama's skirts or papa's legs while they drink coffee, remind one of Algiers, Dakar, Casablanca.

Public opinion has been defined by an American social scientist as "those opinions which governments find it prudent to heed." In this sense, there has been little public opinion in Martinique, where some ameliorations of control and representation were instituted by the French in 1946, after the convulsions of the war, but now the *presence française* insures an opinionless tranquillity that encourages the tourist to tour and the sugercane to bear its sweetness. Compared with the chaotic desperation and color of Haiti, life here is calm and ordered. Independence did not bring prosperity to Haiti: It brought a history of provisional leaders who provisionally pillage and assassinate, and yet the boiling vitality of the Haitian people is evident from arrival. The vitality of the people of Martinique—very similar in their racial composition and history until 1804—is not so visible. There are writers and artists and politicians. But the sense of being "little Frenchmen"—with all the implied advantages and degradations—has made this seem like a toy country, a model form, rather than the real and rooted place which Haiti, despite its misery, manages to be. Haiti is, in Henry James's phrase, a buzzing, booming resort and plantation. It is a curious difference, and while the price Haiti pays is certainly excessive, Martinique has also paid a price for its easier times.

In Europe and America, one carves out a career—a little like carving a roast. In the Antilles, one does not live this way. One chops it up and eats what is necessary, what one can find, day by day. There is not much choice in the matter on a poor island, on an island whose riches are held so tightly. Martinique knows that its long and secret history has barely begun. [1982]

Europe

Ireland: Into the Mystic

JESSICA MAXWELL

THERE WERE THREE OF US: Lucy Brown, me and Lucy Brown's hair, which hangs down her back like four feet of silken sheet metal. It is, we reckoned, about seven years old, and like all seven year olds, it has a boundless capacity for making friends, reaching out in all directions, wrapping its tiny fingers around the arms of perfect strangers, stewardesses, my lunch. It is a mixed blessing to fly 6,000 miles with such a companion, and by the end of the flight Lucy sent her hair to its room, gathering it into a quiet place at the back of her neck and allowing me, for the first time, to see out the plane window. There beneath us, boiling like slate stew, was the Irish Sea. Just beyond that, reaching toward America, the green geometry of its farms spreading out like a flattened, faceted emerald, was the dear and dangerous country of Ireland.

The moment we stepped off the plane we learned that Ireland is a place of the elements. It does not snap a movable metal hallway onto the door of your plane to deliver you from one artificial environment into another without so much as a whiff of the place you've just landed in. When you get off the plane in Dublin, you get off into the wind. And it was at this point that Lucy's hair decided to go home. It quickly turned to get back on the airplane, blinding me in the process so that I did not see the smoky figure waiting for us just inside the airport door.

"Philip!" Lucy yelled.

There he was. In all his rumpled Irish glory. A fine wool tweed suit that hadn't been pressed in a couple of years, a somewhat baffled tie and elegantly scuffed shoes, pink cheeks, sad moon eyes and black hair that obviously would rather have stayed in bed. He looked like a cross between Robert Mitchum and an owl.

He is, in fact, an architect. During Lucy's previous European tour, she had helped Philip with some last-minute architecture assignments in London and they became fast friends. They also became sort of transcontinental soul mates, mainly because Lucy is a collector of all things strange and artful, and Philip is one of Ireland's more entertaining specimens.

"I got your letter," he said in a tone that seemed not only to explain his presence at the airport at that particular moment, but the very existence of time, space and the universe itself. It was difficult to decide if Philip operated in an extremely high state of consciousness or was simply from another planet, but whatever it was it infused the air around him with some sort of unintentional mystery, and when he suggested tea, Lucy and I agreed with near silly enthusiasm. And we soon found ourselves in a car that matched Philip's suit, zipping through Dublin's kamikaze traffic toward something or someone called Bewley's.

"Jersey milk from our own farms," the sign read. The milk was yellow with cream. "Irish brown bread," read another. The slices were thick with oats and wheat. To the back of the building, bevies of women in brown uniforms floated in and out of the steam of some sort of colossal tea-making machine. And the place rang with the porcelain music of happy restaurants.

Quite simply, Bewley's was the most ancient cafeteria Lucy or I had ever seen. All of its parts were framed in dark wood and there was a kind of film over everything. It reminded me of the kitchens of Gulf of Mexico towns, where the atmosphere is laminated with the grease of a thousand fried shrimp and salted humidity. At Bewley's the atmosphere was more bacon fat and tea steam, but the feeling was the same: These are places built by and for the people who live there. Things are simply kept comfortable. And for those of us who live in the cold heart of American quick-change artistry, this is a major blessing. Muscle by muscle, nerve by nerve, we felt ourselves relax.

Between Bewley's balm and basic jet lag, we were practically in a deep trance by the time Philip dropped us off at our hotel. My notes at that point look like they'd spent the last five nights in a pub. All I can make out is that Philip was picking us up again at eight that night and something about Ireland having "done away with animal glue," which I hope had something to do with buying postcard stamps.

Jury's Hotel is one of the better hotels in Dublin. They have tried to make it fancy. They have tried to make it sophisticated. They have tried, actually, to make it American modern. Fortunately, they have failed. You cannot make a sow's ear out of a silk purse in Ireland. And Jury's remains a most homey and comforting big hotel. The bedspreads are soft, the sheets are cotton, the wallpaper's a bit faded, there are rust stains in the wonderful, old, deep, long bathtub, and it doesn't smell like Lysol. But mostly it's due to the help: They are Irish. And as the British should have learned centuries ago, the Irish bow down to no one. There is none of that European servility designed to make patrons feel important, to which my personal response is always an uncontrollable urge to run out and buy the maid flowers. But, being Irish, they are also all heart, so if you do ask for something, they accommodate you with genuine care, as if you were a guest in their own home. For instance, when we asked for Irish brown bread instead of croissants for breakfast, the cook actually sent the waiter to a nearby bakery.

Welsh rugby players, on the other hand, are a different story. They are, in fact, a different species. And unbeknownst to us, they had taken over Jury's for the upcoming Wales versus Ireland rugby match.

We had told Philip we'd meet him in the bar, which is, of course, the natural habitat of all breeds of *Players rugbyis*. Our unexpected entrance sent up a territorial squawk and two silver-tongued crested warblers were dispatched to investigate. They offered to buy us "a pint," referring to Guinness, the black brew that constitutes the major portion of their diet. We declined. They complimented Lucy on her plumage and it responded by dragging itself across their beer. They asked us where we had migrated from and when Lucy said "California," I added that "we have whales off our coast too," thus ending our welcome forever.

IF IT CAN BE ARRANGED, ONE SHOULD ALWAYS TOUR A NEW CITY WITH AN ARchitect. If it can also be arranged, the tour should be conducted during the daytime. Now, Philip is an excellent tour guide; like all good architects, he knows the important buildings by heart, which he dutifully pointed out to us one by one, but since it was dark we couldn't see anything. However, our overall impression was this: Dublin is in a rather deplorable state of decay due to the decline of the Irish pound, inflation, unemployment, lack of import commodities, the price of Guinness. It is, however, also experiencing a structural renaissance, and in the quiet darkness spinning by our car window, piles of 200-year-old bricks and glass sat patiently beside gloriously refurbished historic monuments with newly polished brass doorknockers and brightly painted windows. It was with this sense of restoration that we arrived at a club called Baggots, which Philip himself had recently redesigned, to hear one of the most innovative and beloved rock bands in Ireland, Moving Hearts.

Philip knew half the people there, including Tony, the owner, who is considered the "Pita Bread King" of Dublin, and, true to form, he instantly handed us a couple of overweight pita sandwiches exploding with sprouts and veggies.

Before Lucy and I could digest the significance of eating such utterly California cuisine for our first Irish supper, Moving Hearts moved. Rather they began to breathe, since their songs congregate around the strange, windy instruments of traditional Irish music. Uilleann pipes and low whistles came to life in high-pitched cadences. A bodhran drum created the heartbeat, and a bouzouki wove a string netting around the whole thing with metallic, mandolinlike grace. Suddenly everyone was on their feet. All around us people were bouncing. This was rock and roll?

Clearly, it was more of a jig. Then a synthesizer took over and an electric bass added a mean undertow and almost imperceptibly the jig slid into jazz. And we began to understand. This was something new. It carried with it the raw force of synthesis and supersedure; of colliding cultures and agreement. We decided right there it could only be called Irish fusion.

Then Moving Hearts began to speak, and that was something else again. It was a good lesson. We were not prepared, had not been prepared since the sixties, for what we were hearing. It was passionate. It was political. It was art. It cared.

Remember the brave ones when the
button is down
In an office in Moscow or in
Washington,
And the faceless features of a
child unborn
To a civilization that wouldn't learn
to live

We were dumbfounded. Perhaps sticking up for themselves is simply a way of life, but whatever the cause, when the Irish see injustice, they fight it. They are, above all, a people's people, and right now, this very second, the one thing they cannot understand is America's incredible passivity about nuclear war. It came up again and again. At parties, while shopping, in chance conversations. It makes them feel helpless. It makes them furious. It puts tears of frustration and embarrassment in journalists' eyes and makes us make promises to deal with the subject—at least in the one place where we have some control: our stories. So in honor of Moving Hearts and the fighting Irish, for all the people who fed us and brought us pints and told us tales and played us music and taught us songs, I'm going to stop this story right in the middle to bring you a message:

Dear America,

It has been brought to our attention that you now have the ability to blow up the world. We love our country very much and we would rather keep it in one piece. Do you think you could find time to write a letter or vote for someone who's against the arms race? We would really appreciate it.

Sincerely,
Ireland

WHEN YOU THINK ABOUT SHOPPING IN IRELAND, YOU USUALLY THINK ABOUT buying Irish linen and Irish crystal and Irish whiskey and those wonderful big hand-knit Irish woolen sweaters. What you don't think about is that Ireland is part of Europe and therefore is privy to European fashion. And since the pound was worth about two and a half times less than it was when I was in Europe ten years ago, Lucy and I found that we could buy Italian boots and French dresses for the price of American polyester sweatshirts. And that is exactly what we did the next morning while Philip finished up some work.

As Irish as it is, Dublin remains the cosmopolitan city of Ireland. There are streets and streets of boutiques and good department stores in the heart of the commercial district. But the best part is that it's all so walkable. Everybody walks and shops in Dublin. And even though Dublin is a big city and all big cities take their toll on the human psyche, even though times are tough and people's faces are often policed by private pain as they tense their bodies against the cold, they're all in this together and they know it. You can feel it on the streets. The purpose and strength of it is contagious. It makes you feel powerful. It makes you defiant. It made Lucy and me take enormous fashion risks: Lucy paid $20 for some flat, round-toed shoes that were as green as the Irish countryside; I paid $50 for a Danish jumpsuit with an unusually large belt that gave me a sort of Celtic warrior look. Luckily, just before we started resembling a two-woman chorus line for Boy George, it was time to meet Philip back at Jury's.

Philip does not live in Dublin. He lives in a sweet little farmhouse about 20 miles south of Dublin in County Wicklow, one of the loveliest regions of mountain and lake in Ireland.

Philip led and we followed in our silver, rented "Dan Dooley Knock-

Along" Toyota. Like the English, the Irish drive on the wrong side of the street. They also have their steering wheels and gear shifts on the wrong side of the car. Since I was born with what I call directional dyslexia—that is, I get my right and left mixed up even at home—I figured it would be far safer if Lucy drove. But that was before I realized that the front seat passenger sat face-to-face with the oncoming traffic, and since Lucy was used to driving on the other side, she tended to cut it a little close. She was also trying to follow Philip, whose driving motto seemed to be "born to die," so I cannot say that the journey out of Dublin and into Wicklow was particularly educational. It was, however, a regular Nautilus workout for my knuckles, and once we found ourselves surrounded by the rich and muscular landscape of the Wicklow valley, it was merely a matter of scraping me off the ceiling so I could get out and have a better view.

"These are the Featherbeds," Philip yelled into the wind.

I looked around thankfully for a place to lie down, but there were only great sloping slabs of land that, at certain intervals, had been curiously cut away to reveal the blackest earth I had ever seen.

"That's turf," Philip cried. "What we use in the fire."

Pointing to a big nub of a mountain way off in the distance, he added, "That's Sugarloaf. And this is gorse." He plucked a valiant little yellow flower from a low, wind-battered plant. Then, to my horror, Philip squished the flower between his fingers and stuck it under my nose, and to my astonishment it smelled just like coconut.

"It smells like coconut," I said.

"Yes," he replied and returned to his car.

Lucy was glad to follow suit, having spent the last five minutes chasing her hair, which was chasing itself all over the place in a frantic attempt to get back into the car. And we were off again.

The Wicklow valley is actually one of Ireland's geographic treasures. It has, for instance, the greatest granitelands in the British Isles—molten and angry rock that rose from the viscera of the earth about 500 million years ago and put a permanent kink in the neck of the valley's thick, slatey upper crust. The wicked Wicklow wind further assaulted the poor slate until its health failed altogether. Its powdery remains are now scattered at the feet of its predecessors, the victorious granite mountains of County Wicklow.

It is, perhaps, the ossified influence of this post-Paleolithic panorama that makes Wicklow residents so hardheaded. No sooner had we driven a scant quarter mile than Philip pulled over again, this time to point out an old stone structure down the hill to the left.

"It's a reform school," he explained. "Run by a young couple. That's where they put the bad boys of Belfast."

Belfast.

How was I to write about Ireland without mention of the northern part of the island? The chicken-livered part of me that couldn't handle driving in Dublin wasn't about to risk its life over incurable religious mania. But my mercilessly macabre, maddeningly curious, black cat streak—which led me into journalism in the first place—purred. I tried to give the cat some coconut gorse to play with, but it was no use. Visions of intrigue and obsession seduced my brain, and the gorse blossom fell back into the brittle nest of bracken and bog.

The hills around Philip's house have, if Ireland will forgive me, that lolling, bucolic beauty of a New Zealand travel poster. They are, however, latticed with decidedly Irish stone fences made hundreds of years ago with the stones gathered when the land was originally cleared. The reason they haven't fallen down yet is that they were stacked with spaces between them—a sort of stone crochet—so that the wind could get through. From a distance the fences look like draped necklaces that surround square, whitewashed farmhouses, set jewellike and apart on the green throat of the land.

It is a vista of extravagant beauty. But you have only to lift your eyes to see a sight that easily makes you forget to breathe: the Wicklow cloudworks.

Their velocity is phenomenal; their antics endless. Huge and luminous, they tumble like cauliflowers boiling in the bluest of soups.

Philip shares his rented 300-year-old, three-bedroom, $150-a-month farmhouse with Cyril, an award-winning photojournalist. And they share it with all their friends, young people like them, mostly in their thirties and college educated, who work in Dublin but live, on principle, in the country. They move freely, stopping by for tea, dropping in for dinner, borrowing tools, bringing by a new album, in a way that quickly flushed out Lucy's and my own American territoriality. We are far more distanced than they are, even from our close friends. We draw much clearer lines between ourselves, our possessions, our time, our space, our futures. After living one week among Philip's friends, I was convinced that we are suffering terribly from what can only be called a loss of tribe.

You never know who you're going to have breakfast with at Philip's, and when we walked downstairs that first morning, two strangers were having tea at the kitchen table. One was a carrot-haired man, the other was a tall woman with an Irish accent that rivaled any we'd heard so far. It was a delicate thing that rose at the end of each sentence like a question, and her pronunciation was strangely soft. She would, for instance, say "th" as a "d" or a "t," so that "the" came out as "da" and "thing" sounded like "ting."

Lucy and I made our somewhat self-conscious entrance.

"Ah, the enlightened women," the man said. "And should I be afraid of you?" He smiled an extremely crinkly smile and dutifully poured us each a cup of tea, one of which was intercepted by Cyril, who had just cruised through the kitchen like a fast-motion Groucho Marx.

"Breakfast for four? Oh, eggsellant," he said in a voice that sounded like a cross between a horse's neigh and Grandma Moses. "Would you have an egg? Or will it be porridge now?"

"Sounds like you're on your last eggs, mate," the red-haired man said.

"And you can go to shell, you can," Cyril replied, and plugged in the electric kettle for more tea. Then he disappeared into the living room, and in a few moments the most joyful Irish music filled the house.

"What is that?" Lucy asked.

"Why, it's the 'Song for Ireland,' it 'tis, it 'tis," Cyril said, "and a mighty tune, that, a might-tee tune."

"It's by De Danann," the woman explained. "They're a favorite traditional Irish band."

"Don't you think you should introduce us," the man said to Cyril, who was now poking pieces of black peat into the wood-burning kitchen stove. They filled the house with a delicious burnt-leaves scent that clung to our sweaters like Irish perfume.

Cyril whipped around and, with a courtly bow, he said, "Grainne, Lucy, Barry, Lucy, Grainne, Jessie, Barry, Jessie, have another cup of tea?"

"Well, it has to be made with completely boiling water," said the man, who seemed to be named Barry.

"Aughk, brother!" said Cyril.

"You're going to learn something here, Cyril. Tea must be made with boiling water. Coffee must be made at just under boiling water."

"And tea goes much better with Irish brown bread," Lucy added.

"Are we out?" Cyril asked with extreme mortification. "Just a minute, then," and he really did begin to neigh like a horse and make running hoof sounds across the kitchen floor.

"Would you like some hoof and hoof in your tea?" asked Barry.

"Oh, Barry, they just woke up," laughed Grainne.

The music from the living room spun into a spirited jig, and Lucy asked Cyril, "How do you dance to this?"

"You'll see when we get to Galway," replied Philip, who had just slinked into the kitchen, his hair still wet from a bath.

"There are jigs and reels," Barry said. "Cyril, what's the difference between a jig and a reel?"

"On one you catch trout, on the other you catch salmon."

"That reminds me," said Philip. "Maybe we should take you to the Aran Islands to dance."

"No, they'd be taken off the floor," Barry replied.

"What floor?" I asked.

"There would be strange men with big arms and big hands and they dance with these girls and just lift them off the floor," he said and flew an imaginary partner above our heads.

"Yeah, they don't know their own strength," Philip explained. "They're all fishermen. They row all the time."

"Yes, they do have a lot of rows," Cyril added.

"Sometimes they have salmon roe," Barry added.

Grainne rolled her eyes.

"Yes, they don't know their own strength and, of course, they mightn't be the full shilling either."

"Do you mean they might not be playing with a full deck?" Lucy asked.

"Eggsactly," Barry replied.

And so it was decided that we should go to Galway on the west coast of Ireland. And it was decided that Philip would go with us, if we could wait a few days for him to finish a project. It was also decided that Cyril and Philip were late for work and that Grainne and Barry would take us for a drive around Wicklow that morning. But not before Grainne went home to take some soup off the stove. Lucy went upstairs to brush her teeth, I poured another cup of tea, and Barry began thumbing through a book I'd left on the table the night before.

"*The Tao of Pooh*?" he said. "So Winnie the Pooh's a Taoist? Now that's interesting. Do you think he is?"

"Well," I began slowly, "the author, Benjamin Hoff, thinks that Pooh is a, well, a Western Taoist master."

Barry opened the book and started reading out loud from the preface:

"What's that?" the Unbeliever asked.

"Wisdom from a Western Taoist," I said.

"It sounds like something from *Winnie-the-Pooh*," he said.

"That's not about Taoism," he said. . . . "It's about this dumpy little bear that wanders around asking silly questions, and making up songs, and going through all kinds of adventures, without ever accumulating any amount of intellectual knowledge or losing his simpleminded sort of happiness. *That's* what it's about . . . "

"Same thing," I said.

BARRY LAUGHED. "THAT'S WHY I LEFT TRINITY AND WENT TO PERU," HE SAID.

"Trinity College?" I asked.

"Trinity College," Barry replied. "In Dublin. I was a math professor there."

"So what are you doing now?" I asked.

"Whatever needs doing," he replied.

"And where do you stay?"

"Wherever I can."

"Should I take the camera?" asked Lucy, who had just come down the stairs in a cloud of spearmint toothpaste. I was pleased to see that her hair was contained in two neat braids. Grainne arrived moments later, and we all piled into our silver Toyota Knock-Along.

Lucy turned on the radio and a news broadcast came on—in Gaelic. It sounded like a cross between German and Hawaiian.

"Did you all study Gaelic in school?" I asked.

"Well, I could have," Barry replied, "but I'm from the Dark North, you see. British occupation."

The Dark North. The cat's tail twitched.

"What city?" I asked.

"Belfast," he said.

Belfast. The cat pounced.

"Barry . . . "

For the first time, Barry's elfin smile faded and an almost painful weariness replaced it.

"I don't like to talk about it," he said. Then he looked at me and he took a deep breath.

"Okay," he said, like an old man who had just agreed to let a dentist pull all his teeth.

"If you were to drive up from Dublin to Belfast, you would cross the border just after Dundalk and into a town called Newry. And at the border you will generally notice a lot of British Army land rovers, jeeps and small armored cars or what they call armored personnel carriers, and going through the center of Newry, you will notice that the police stations and public buildings have wire grilles all around them and are all fenced off, and this is to stop people from throwing petrol bombs into them. Sometimes the buildings will be inside a huge, metal box. This is to stop people from shooting into them.

"Then as you go past Newry, you go onto an actual motorway, like a freeway, which brings you into Belfast. And when you arrive in Belfast, the first thing you'll probably see is smoke. It's kind of a dirty city. But if you're to go into the center of the town, at some point you have to pass through a security gate in order to get into the shopping area. And the people have got kind of used to this business of being searched every time they go into town.

"Now you can do a little tour and go off into places like Bally Murphy and the Falls Road, which are the Catholic ghettos and are in terrible condition. They're a shambles, a lot of homes have been burned down. There are few facilities. Then the Protestant side of things is the Shankill Road, which is a couple of streets away and looks very similar. And between the Falls and the Shankill, there's what they call a Peace Line, which is a long, huge wall that they built all the way along between the two.

"And most of the people go on about their business and it is possible to lead quite a normal life and the only real inconvenience would be the fact that your car may be stopped and searched at any time, or you may personally be searched by the British Army or the police or what they call the UDR, which is the Ulster Defense Regiment, a part of the British Army based in Northern Ireland.

"Most people can go up and stay a week or two quite safely because they'd always be taken out by people who know where they should and shouldn't go.

"The most dangerous places are the border towns, which are villages that are very strongly one way or the other, and all you'd need to do, as they say up there, is kick with the wrong foot in the wrong town."

Barry leaned back against the seat and lit a cigarette.

"You know, it's supposed to be a revolutionary struggle type thing, or a struggle for a united Ireland, but it has turned into a sectarian shoot-out between the IRA and the Protestant organizations, and that's all there is to it."

"TURN DOWN THERE," GRAINNE SAID SUDDENLY, AND WE FOUND OURSELVES heading downhill toward what looked like a park beneath a mountain. In its middle was a huge brown lake.

County Wicklow is known for its lakes, which filled up with water after the glaciers scooped holes out of solid rock like so much granite ice cream. This one, Grainne informed us, belonged to Garrett Brown, a member of the Guinness family.

"I talked to him all afternoon once," Grainne said. "He's got a lot of wild stories and all that crack."

We parked by the brown lake and walked over to its shore and just when I was going to ask Grainne if it was full of Guinness, Barry started splashing around in his rubber boots. "These are called Wellingtons, you know," he said. Then he began to sing:

Oh Wellies they are wonderful.
Wellies they are swell.
'Cause they keep out the water

and they keep in the smell.
And if you're in a room of folk
You can always tell
When someone takes off their Wellies.

"By George," Lucy said. "I think we've found our Irish Pooh!"

"Yes!" I agreed. "And we'll have to call him Beary."

We were on the road again, high above the estate and all around us were coarse, mauve-colored bushes.

"That's all heather," said Grainne, one of the best bog guides around. "And those are Irish blueberries. They're called frockens. And there's some bog cotton. And you see how all the water here is tea-colored? Well that's because of the tannin, sort of a bog tea you could say."

It was getting dark. On the way back we stopped and bought things for dinner, which was a good lesson in Irish imports. There were red yams from Egypt, green beans from Kenya, cabbages from Holland and fennel root from France. There were also, Lucy and I noted, bottles of Mr. Clean, which, in Ireland, is called "Mr. Proper," some heavy-duty cleanser called "Gumption," and little cubes of beef bouillon called "Oxo," which was guaranteed to have "more man appeal."

Everyone showed up for dinner that night. Barry and Grainne, Philip and Cyril, Lucy and me, and we discussed our trip to Galway. Philip decided that we ought to overnight at his parents' cabin on Lough Conn in County Mayo, just north of Galway. There was a reason for this: Philip had become disenchanted with the life of an architect and was seriously considering the advantages of a life as either a musician or a furniture maker. He had already begun to collect flutes and whistles and odd pieces of wood, and there was an elderly furniture maker in County Sligo, not too far from the cabin, whom he wanted to visit.

That night, as Lucy and I wrapped ourselves around our hot water bottles, the wicked Wicklow wind shipped up a mournful lullaby, and from somewhere in the night we heard Philip answer with a wheezy, uncertain Irish melody on his new tin whistle.

AFTER A BREAKFAST OF PORRIDGE, SAUSAGES, BACON AND TEA, WE TOOK OFF in a northwesterly direction. Lucy drove, Philip navigated, I prayed and Barry went along for the ride. Eventually we crossed the Shannon River, which ran right across the road. It's a boggy thing, full of estuaries and reedy weeds, and in every little river town we passed, the schoolchildren had their Wellies on.

John Surlis lives with his wife in a simple house in a village called Ballaghaderreen. All his life, he has made his living making one thing: the Irish kitchen chair.

Philip had brought him some apple wood. He hoped to get a lesson in exchange and he did. Mr. Surlis walked from behind the counter of his little store into his icy workshop and sat down on a benchlike contraption called a gray mare. His chickens watched him through the window.

"We have six or seven generations of furniture men in the family," he said. "We can trace it back 150 years."

He clamped a stick into the wooden jaws of the gray mare and began to work it with an old metal plane that almost looked cruel in his arthritic fingers.

Scrape. Scrape. Turn. Scrape. Scrape. Turn.

"I used to make everything in wood. Them tubs there. Milk churns. We made the ones that brought the butter to England in olden times."

Scrape. Scrape. Turn. Scrape. Scrape. Turn.

"We made the cradle. The old-time coffin. Some of my chairs even went to California."

Slowly, slowly, the stick moved in his hands. Slowly, it became a perfect chair leg. This, I thought, is a man who knows how to live.

Before we left, Mr. Surlis asked us to guess his age, and Lucy did.

"Eighty-one," she said.

"I'll be 82 this year."

"And what," I asked, "is the secret of your long life?"

John Surlis didn't hesitate.

"Work hard," he replied, "and never pretend you're getting old."

PHILIP'S FAMILY COTTAGE, ON THE SHORE OF LOUGH CONN, IS NEXT TO A SECRET all-night pub. The night we were there was special because it was the last night that Sean Kane would own it.

There was a furious turf fire going, and the local folk were gathered around it. Ireland's blood-chilling climate does that to people; they need the fire of alcohol and cigarettes, of pink salmon and red meat, of fast music and breathing bodies, just to keep warm.

For hours we drank Guinness and laughed and met people's relatives and watched the police chief get drunker and drunker until he was jumping over the bar to help Sean pull pints. When we finally made our way back outside and staggered up to the cabin, Philip suddenly leaned over and, grinning his otherworldly grin, said, "Sean says he can get us some poitin."

Sure enough, the bottle was delivered to our door the next morning, wrapped in newspaper.

Poitin is homemade Irish potato moonshine, and rumor has it that a bad batch can blind you, although neither Barry nor Philip knew anyone who'd actually been blinded. There is, however, what is known as the milk test, wherein you drip a drop of poitin into a plate of milk and watch closely. If it curdles around the edges, it's bad stuff. Ours didn't curdle. But it sent such powerful shock waves across the surface of the milk that I was filled with apprehension.

Barry and Philip drank theirs straight. Lucy sipped hers carefully. I wouldn't drink any at all, until Barry made me a poitin toddy with lemon and honey, which tasted like warm vodka.

After our morning cocktail, we repacked the car and headed down the road. We made several stops on the way to Galway. One was at Philip's favorite deserted castle, which featured a spiral stone staircase that rose straight up four stories like a twisted spine and deposited us at the threshold of a missing wall, which offered a thrilling view of the green, green flatlands of western Ireland.

"The teddy bear's paws," Philip said.

"What?" I asked.

"The teddy bear's paws," he repeated. "You know, if you look on a map, Ireland looks like a teddy bear reaching toward America. The paws are County Mayo."

"And his brain's Northern Ireland?" Barry asked with a raised eyebrow.

"And Dublin's on his back," Philip laughed.

"And Belfast is on his neck, where he gets his ire up," Barry added.

"Or his IRA," Lucy said.

They each took a swig of poitin. Down below, a horse rolled in the mud.

The other stop we made was for cigarettes in a shop called Dunny's in the tiny village of Finny, or it might have been the shop called Finny's in the tiny village of Dunny, but anyway, the woman who ran it was utterly charming.

"Are you on holiday?" she asked in a bright, birdlike voice.

"No, we're on assignment," Lucy replied.

"Same thing," she said.

The rest of the drive to Galway belonged to Van Morrison and the Irish paintbox goddess. While his passionate "Into the Mystic" rolled out of the radio, colors rushed and blurred around our car windows. Toasted colors. Treacle colors. Mauve and hay, chartreuse and cream. Peach dust, burnt rose, powdered tea, pale lizard. And every so often, a dear little Irish farmhouse nestled into this renegade rainbow-like blank spaces on the canvas.

By the time we reached Galway, the goddess had painted the day blue. It's a seacoast town, full of the cool vapors of the wild Atlantic. But it is cozy and small, just like Freddie's house, which served as our home base for the whole visit.

She is a marine biologist and very Irish. She is tough and sweet, independent and knows how to be a very good friend. It was Freddie who took us to the best sweater shop in Ireland.

It was run by a plump woman with rude red lipstick painted far above and beyond her lips, and a strange, shrill voice that was forever chattering to the handful of motley men who seemed to just hang around the shop and stare at things.

"They're a bit daft," Freddie warned us. "And prepare yourself for the odor."

We were not prepared. Apparently the owner's dogs and ensuing puppies were given free rein of the place without a convenient exit to the great outdoors. Lucy and I bravely complimented the cute puppies, raved about the stock and were invited to sort through the warehouse of hand-knit Irish woolen sweaters, which sold for about $32 each.

There was another communal supper that night at Freddie's, with the usual comings and goings of old pals.

After dinner Freddie piled us into the old Knock-Along and we drove to a pub famous for its "sessions." Musicians from all around came with their fiddles and whistles and pipes and flutes and beautiful, clear voices and played nonstop while the rest of us pushed in around them like ants.

The party did not stop when the pub closed. It simply moved to the youth hostel across the street. And that's where the dancing began.

The musicians realigned themselves and squares of couples began to form. A jig exploded from the fiddles and the couples began to jump, making an enormous, rhythmic racket with the heels of their shoes.

"That's called battering," Freddie explained.

Lucy, Philip, Barry and I decided to try it, but between the poitin and a distinct lack of skill, we ended up sounding more like a stampede. Then Philip started dancing with Barry, Fred Astaire-style, his cape flying. Barry handed Philip to Lucy and her hair, which had already started dancing behind her back. Then he slid over to me and, together, we did a dumpy little jig around the room. [1984]

A Landscape to Cherish: Scotland's Shiant Islands

ADAM NICOLSON

I HAVE A HABIT IN BOOKSHOPS, partly sentimental, partly pedantic, of which I am not especially proud. I go to the atlas section. I take a book, flick to the page for northern Scotland, and run my finger up the jagged, notched outline of the north-western coast, where a narrow sleeve of water, the Minch, divides the inner from the Outer Hebrides. If three little triangular pimples appear on that expanse of sea between Skye and Harris, I am happy. If not, the map is worthless, the cartographers indolent, and the atlas, for all its representations of freeways and megalopolises, a waste of time. Those three pimples are the Shiant Islands, which belonged to my father and, for the time being, belong to me. The name means "holy" or "enchanted" in Gaelic, and they are the place I love best in the world.

Each of them is about a mile long and half a mile wide, with a scattered fringe of smaller islets and rocks around their coasts. In all they add up to no more than 500 acres, a stretch or two of rough grazing, enough only for a couple of hundred sheep, which one would pass by in a few minutes if these specks of land were buried deep inside the mainland of Scotland. But they are not. They are splendid, tall, proud things, out on their own in the Minch, rising 600 feet out of the sea in sheer black basalt cliffs and surrounded by some of the roughest waters in the British Isles.

Even on the calmest days, every turn of the tide cuts up the surface of the sea here into a turbulent and broken mass of water, a danger for shipping ever since people have inhabited the Hebrides. The Shiants are guarded by their own sea spirits, the Blue Men of the Minch, wild storm kelpies with long blue hair, who board any boat trying to make its way down the Sound of Shiant in bad weather, and quote at the captain the first line of a Gaelic lament. If the captain is able to continue their song, with its difficult schemes of meter and rhyme, the ship is saved. If not, he and his crew are drowned, and the ship lost.

No one lives on the Shiants now, but the memory of people who lived there in the past is everywhere. Apart from a break in the middle of the 19th century, the islands had been continuously inhabited since long before their written history. A Celtic hermit, one of those strict and romantic churchmen who abandoned society and sought God in lonely places, built his cell here, perhaps in the ninth or tenth century. The tumbled stones of the cell are still there, nosed at by the sheep, surrounded by the twittering and fluting of the snipe in the marsh.

The medieval history of the islands is a blank. But as soon as they appear in the records in the 16th century, the Shiants are full of people. There was a small church on one island (the graveyard was still marked on maps a century ago), and the accounts of visitors talk of rich grazing for cows and sheep, of hayfields and even cornfields. Six or seven families probably lived on the islands until the early 19th century, and the remains of their cottages and crofts and of the stone walls that divided their properties can all still be found, crumbled and lichened, but articulate enough.

A disaster occurred at some time in the 1780s: Three members of one family were chasing sheep, and fell over the cliffs and died. By about 1815 the islands seem to have been deserted. The people may have been cleared off by the landlords in a small example of the evictions that occurred all over Scotland at that time, to make way for a profitable sheep farm. Only in the 1860s was a tenant reintroduced, and his children, the last permanent inhabitants, left in 1911, finally driven away by the isolation of Shiant life (two incestuous children had been born in the previous 20 years) and the rigor of the long, dark winters.

The real presence of these people on the islands, the presence of their memory, is important to me. They are part of the meaning of the place. For all the isolation and wild removal from the ordinary run of things, the Shiants are not a slice of the wilderness. They are a human landscape, farmed for perhaps a millennium, and they continue to be shaped by the same forces that have always shaped them. The sheep are owned and tended, in four or five yearly visits, by shepherds who live on Harris.

They come over in a fishing boat, lodge in the one small cottage that is still in repair, see to the ewes and the lambs, deliver the rams for their winter tupping, and mend the fences and *fanks*—a Gaelic word for the arrangement of holding pens and sheep dips through which the flock is collected, sorted, and cleaned. All this is a constancy, a continuation of how things have been and how, I hope, they always will be.

There is another, far older permanence here: the permanence of nature itself. The Shiants throb and teem with life. They are one of the most important breeding places for seabirds in the British Isles.

Something like 77,000 pairs of puffins breed here every summer, arriving in ones and twos every April from their winter on the Atlantic to reclaim the burrows they had the year before. Razorbills and murres pack in on the shelves of the basalt cliffs. The oily, deep green shags, ancient birds that have remained unchanged for 60 million years, creep into their hovels in the boulder fields below the cliffs to continue the line, go fishing, and then stand like feathered crucifixes, drying their outstretched wings, on a rock where the wind will catch them. Great skuas coast all day over the heathery moorland on the tops of the islands. Fulmars, ever on the increase, wheel in perfect, nonchalant flight on the cliffside updrafts. The kittiwakes chatter all day on their nests, saying "kittiwake, kittiwake" as if one's own name was all there was to say. Out at sea, the gannets cruise and plunge, even in the fiercest storms, for the fish they can somehow see beneath the surface of the Minch.

But there is more. In springtime the turf is thick with flowers: cotton grass in the boggy peat of the higher places, orchids on the dryer patches next to them, flag irises and water mint in the little streams that run down from the springs to the sea, medallions of rose-pink stonecrop on the boulders, and tufts of thrift and sea campion next to them. Thick jungle patches of ruddy oarweed spread out at low tide in the sheltered bays. On the seaside rocks, the black and ocher lichens continue to grow at a millimeter or two a year.

Sea otters, those most mysterious and hidden creatures, will sometimes come up onto the shingle beaches and eat a fish held between their hands—the strangest mixture, if you see it, of delicacy and savagery, as they chew at the head of a saike or a lythe as though it were asparagus. Hauled up on the shore, never near the otters, the seals laze in the summer sunshine and then plunge off to play and bob in the sunlit pools. If you float in a dinghy patiently for an hour or two without noise or movement near one of their favorite places, they will come near, six or ten feet away, bristly, wide-eyed, and mustached, as curious about you as you are about them, hesitant, intelligent, as quivery in their way as deer.

Around them—you think of it in the boat as you lie waiting in the sun—the seabirds are fishing, their wings beating like oars, and below the deep blue, lobsters are crawling their way toward their antediluvian dinner.

To own all this? Of course not. Ownership is an irrelevance in front of this. At least nominally, in the eyes of the law, the Shiants have been owned this century by strangers: Lord Leverhulme, the Lancashire soap magnate, who thought they might do well as a goat or rabbit farm; Sir Compton Mackenzie, the novelist, whose wife refused to accompany him there; a Colonel Macdonald, who had in mind a stud for racehorses; and my father, who bought them in 1937 when he was 20 for £1,400 after he had seen the disenchanted Colonel Macdonald's advertisement in a newspaper. Fourteen years ago he gave them to me, as I shall give them to my son Thomas, who is now six, when he is eighteen. And so on, I hope, like the shags, generation by generation.

But there is a paradox here. These islands in all their complexity of life are unownable, but they are my islands. I look for them in bookshops. Late at night in the winter, in strange hotels, I listen to the radio for the shipping forecast—"Hebrides, Minches: storm force 10, veering southwesterly, sleet, visibility 200 yards"—and think of them then, wet, battered, impossible. On every flight across the Atlantic, I peer out for them, looking for an opening in the clouds (it has happened, magically, only once) to see them there still and maplike below me, with the sea sheened and glittery around them, while the stewardess hands out headsets and warm towels.

We talk in our family about this owning-but-not-owning, and it has come to a simple formula. We do not own the Shiants; the Shiants own us. Ownership then becomes a matter of duties not rights. If people mention the words "lord of all you survey," I shudder. It is a phrase born in the tyranny of ignorance and in an egotistical contempt for the natural. I feel I have no rights over the Shiants and only one duty: to look after them, which should perhaps be seen as a right in itself. The law is my greatest help in this. The Nature Conservancy Council, an agency of the British government, has designated them a Site of Special Scientific Interest, principally because of the birds and the unique geology of the islands, which is a series of Tertiary basaltic sills, the outpourings of a volcano in Skye about 50 million years ago. Under this designation I am not allowed by law to do anything to affect the natural regime of the islands—no additional structures, no fertilizing of the grass, no introduction of alien species, no increased access, no hunting. There is no jetty—one can only land in good weather on the storm beach—and I am not allowed to put one in. The small house is a rough and ready thing; I could not, even if I wanted to, make it larger or more luxurious. There is no electricity, telephone, running water, or lavatory. Nor will there ever be. The Shiants are to be kept as they are, inviolate and protected.

It is, in a way, a strange form of care by denial, an almost purely negative way of guaranteeing richness. But that paradox disappears in the presence of the islands themselves. There is a wholeness there to which, in truth, I do not belong. The only real threat to them lies in me, or in someone like me, who might use them harder, improve them, and distort them. They have their own logic, their own successes and disasters, their own way of being, which runs unknowably deep.

The best thing I can do for the Shiants is to give them a sort of love by absence, a visit now and then to see that they are what they are, an occasional longing look from an airplane window many thousands of feet above them. [1991]

Budapest's Margaret Island

JOHN McKINNEY

A HUNGARIAN GYMNAST pumps iron in a Jacuzzi. A German businessman eases his angst with an underwater massage. A rotund human of indeterminate sex lies packed in mud like a happy hippopotamus. Grunts and groans, oohs and aahs, and gerontological gurglings of delight mix with the vapors rising above the tubs and pools of the Thermal Spa. They have come, in all shapes and sizes, from East and West Europe, to this small island in the Danube to take the cure.

"Cure is perhaps too strong a word," cautions Dr. Edgar Hotchkiss, medical director of the Thermal Spa, as he makes his morning rounds through the treatment rooms. "But for ailments like arthritis, rheumatism, and various locomotor problems, our water therapy has lasting benefits for three to six months, even a year."

Apparently, many Europeans and a growing number of North Americans agree with Hotchkiss. The island's hotels are filled with spa-goers, though most would seem to fall into the category of the Worried Well rather than the Seriously Ill.

As Hotchkiss makes his rounds accompanied by the cheerful Nurse Eve Hamburger, he exhibits a kindly and convincing bedside—or more accurately, poolside—manner. Why else would someone on vacation agree to be pummeled, parboiled, and packed in mud?

"North Americans just don't understand the value of balneotherapy," he comments, reading my mind. "They think it's a cult or quackery or something only for old people."

Hotchkiss is a striking figure, one of the world's leading authorities on balneology. A thick shock of salt and pepper hair crowns a visage that belongs more to an artist than a man of science. In the right light, he could double for Franz Liszt. He directs a symphony of water treatments, all of which are made possible by the island's natural hot springs. The medicinal waters enter the spa's hydrotherapy equipment at a temperature of 158 degrees F and are cooled down to a tolerable 104 degrees F. The essential treatment element of the water is slowly decomposing sulfate; however, the typical sulfur odor is not present, and the island is not one of those places where one imagines hell has bubbled up. The natural gaseous content of the water is significant. An even film of little bubbles tickles the surface of the body and the effect can be downright sensual, like romping in warm Perrier.

"The sitz baths, the carbon gas baths, and all the other treatments are only half the story," declares Dr. Hotchkiss. "This is the only world-class spa on an island. The environment of the island—the gardens, the river, the tranquillity—can have a powerful healing effect."

I witnessed some of the island's healing power earlier in the week when I met six members of the Bokay family, reunited on the island after a 30-year separation. They seemed to be drifting slowly along the Danube pathway as if in a dream, but this was no dream. Thirty years ago Soviet tanks rolled into Budapest, compelling one side of the Bokay family to flee to Austria and then on to America, while the other side stayed on to witness what is now known, depending on one's political persuasion, as "the invasion" or the "counterrevolutionary uprising." The events of 1956 split this family as surely as this island splits the Danube; one branch of Bokays settled in Szentendre, about 12 miles north of Budapest, and the other branch landed 10 time zones away in San Francisco.

Margaret's very location—floating in the Danube between two parts of Hungary's capital, in a country positioned between East and West—provides an ideal site for a reunion, for a healing of the most important kind. During their week-long reunion the Bokays soak in the island's hot springs, eat too much goulash and strudel, and listen to outdoor concerts given by the Dixieland jazz bands that Hungarians are crazy about. Often they walk along the Danube and talk of many things, of Volgas and Volkswagens, communism and capitalism, Szentendre and San Francisco, of life in the old country and life in the new. But in the first hour of their reunion there is only hugging and holding and tears and good-natured teasing about hair gone gray and bodies gone round and expressions of astonishment at how the years could have passed so quickly.

Politics and economics have, over the past 30 years, changed the face of Budapest. But what has not changed in 30 or even 130 years is the feeling the city evokes among Westerners glimpsing it for the first time. It's an approach at once mysterious, exotic, and surprising, particularly if that first approach is made by river. A busy river it is, crowded with an international brigade of ferry boats and hydrofoils, fishing boats and freight barges. On one side of the Danube are the Buda hills, said by some geologists to be the last gasp of the Alps, rising dramatically from the water. On the other side of the river is Pest, low and flat and stretching into the plains, *puszta*, scorched and silent. Marking this meeting of Europe and Asia are towers and turrets, castles and cathedrals. Chief landmark is the Hungarian House of Parliament, a Gothic giant in limestone that resembles London's Parliament—with the Danube substituted for the Thames. And over on Margaret Island is a tall bronze monument symbolizing the 1873 union of old Buda and new Pest, the result being Budapest, surely a more euphonious name combined than divided.

The train route to Budapest is equally magnificent, recalling the grand old days of railroading when the way to reach the "Queen of the Danube" was via the Orient Express from Paris or the Wiener Walzer from Vienna. The tracks waltz along with the wide bends of the Danube. This approach by rail captivated that great observer of river life, Mark Twain, who visited Budapest in 1899 to deliver a lecture during the Jubilee Celebration of the Freedom of the Hungarian Press. An enterprising cub reporter from the *Journal of Pest* bluffed his way into Twain's compartment for an exclusive interview. Unfortunately, their common tongue was German, a language Twain could barely speak and one he continually lampooned. The young scribe also had difficulty competing for Twain's attention with the panoramic view framed in the train window, as he dutifully reported:

> *I must confess after the Danube Bend, the old man became more and more uncommunicative, and in an appropriate moment escaped into a neighboring coach. I was not hurt by this behavior since it made me feel good to see the great American humorist watch with curious eyes as Budapest unfolded itself to our view. Already around Visegrad, he burst into enthusiastic exclamations and looked with amazement at the immense Danube winding its way into the picturesque mountains with majestic quietness. He became speechless; excited as a child, he gazed through the window.*

Compared to approaching Budapest by boat or train, the drive from the airport has to be rated as something less than inspiring. Budapest's backside is a mélange of dirty brick factories, smoke-belching buses, fatigue-clad Hungarian army recruits laboring on public works projects, and grim Soviet-style concrete block apartment houses—six, eight, ten stories of unreinforced masonry and soulless design.

Aesthetic relief for the motorist comes only when the Danube and its eight beautiful bridges come into view. Oldest of the bridges so beloved by Budapestians is Chain Bridge, built by Englishman Adam Clark in 1849, destroyed by retreating Nazi armies, and rebuilt and reopened in 1949. Some Hungarians rave about the graceful Elizabeth, a suspension bridge, while others praise the Liberty and Petofi bridges. Two bridges, the Arpad and the Margaret, link Margaret Island with Buda and Pest. The Margaret, designed by French engineer Alexandre-Gustave Eiffel, is especially intriguing because it abandons its straight trajectory about midriver and heads off at an obtuse angle. From certain angles the bridge appears to end in midair.

Part of Margaret Island gives the impression of being an open-air gymnasium. The island boasts tennis courts, boat houses, jogging paths, and soccer fields. The red brick building of the National Sports Swimming Pool, designed by Hungarian swimming champion Alfred Hajos, hosts the serious swimmers, while the huge Palatinus Outdoor Swimming Complex with its cold and warm water pools—complete with artificial wave maker—can accommodate up to 20,000 swimmers and sunbathers.

As sunset approaches, thoughts turn from recreation to romance. Couples of all ages stroll the delightful promenades, through woods thickly populated with statues and busts of Hungary's foremost writers, sculptors, and artists. At the Casino, an open-air nightclub, a clean-cut rock band sings "Let's Spend the Night Together" in Hungarian. Young people dance past a wall bearing the only graffiti I saw in Hungary: "Join the Party—the Garden Party." Other couples neck at the open-air cinema and at a Dixieland band concert under the stars. Every park bench is filled with smooching and soft words.

"Margaret Island has always been a romantic place," opines Hungarian octogenarian Eugene Kornfein. "And don't think I'm so old I can't remember romance." Kornfein confesses to having taken a few pretty young ladies for romantic walks in the woods during his younger days. "Margaret has always been the place in Budapest to take your sweetheart," he adds with a big wink.

As twilight falls, Kornfein surveys the island from the steps of the Thermal Hotel. From faraway float the sounds of blackbirds and a gypsy band. "The fine old trees, the rose garden, the Budapest skyline, the river . . . it's all very romantic."

I disagree with Kornfein only about the romance of the river. I have observed the Danube at sunrise and sunset, in sunshine and rain, and seen the river take on many colors—black, slate, gray-green, and dull brown—but never blue, and blue is the proper color for a romantic river.

"Now who told you the Danube is supposed to be blue?" Kornfein asks. "Not a Hungarian. Not a poet or even a journalist—just a Viennese waltz composer."

Kornfein says he first visited the island in 1910. It cost a few pennies to walk across the Arpad Bridge to the island, a charge that kept the poor off Margaret. A rail line with a horse-drawn trolley took visitors across the one-and-one-quarter-mile length of the isle. After the Great War, Kornfein lugged his thick law books over to the island every day and spent his afternoons studying torts in the shaded ruins of Saint Margaret's convent.

Two thousand years ago the Romans built a pontoon bridge from the Pest side of the mainland to the island and built a watchtower to keep an eye out for enemy ships. The commander of the Roman garrison at nearby Aquincum had a summer residence on the island, as did other prominent Romans. Archaeologists have discovered remains of Roman baths and harem quarters and suspect that Roman interest in the island was more sybaritic than strategic.

Margaret Island is named after the only daughter of a 13th-century Hungarian king named Béla IV. Fleeing before a Mongol invasion, the monarch vowed that if God would lift this scourge from his land, he would consecrate his Margaret to God's service. The Mongols were defeated, and a cloister was built on the island to house not only nuns and novices but also the country's foremost school for young ladies. The presence of the princess attracted the daughters of nobility, many of whom were not really interested in saintliness and looked askance at the profound devotion and extreme ascetic practices of Margaret, who took her vows of poverty very seriously and was inclined to hair shirts and charity. Not fond of performing the cloister's menial tasks, the young women reputedly showed their displeasure by dumping buckets of slop water on Margaret.

Princess Margaret was apparently not only saintly, but also beautiful—even in her humble robes. As the comely daughter of a king, she was an eminently desirable match, with several royal suitors after her hand. But Margaret was repelled by the hirsute faces and rude virility of these young men and longed for a quiet gentle man. Alas, a suitable suitor never presented himself, and Margaret died in the convent at the tender age of 29. But Margaret has never left the memories of Hungarians; early this century, nearly 700 years after her death, she was canonized by the Church.

The Turks invaded Hungary in the 16th century and renamed the island "Maiden Island." The Turks liked the fine old buildings and idyllic setting, and it is believed that the island was the site for the pasha's harem. When the Turks were finally defeated, they destroyed all the island's buildings.

A much more engaging occupation, this time by poets and artists, stretched from the middle of the 19th century to World War II. Besides the Grand Hotel, which catered to the rich and noble, there was the more modest "little hotel," a haven to these artistic folk. These poetic souls all fell in love with the island, their green refuge away from the turmoil of the city. Artists and writers are much appreciated in Hungary, and so when these cultural treasures were down and out, a nearly unlimited credit was extended to them. Landlords were amply repaid with sonnets and etchings. One writer, whenever he sold a story, would phone the mainland to order some special beef and have it brought to the island by carriage. He would then prepare a goulash, which was as famous as his writing, and invite all his friends and creditors.

The island has been a celebrated spa since 1866, when deep-drilling operations conducted by Vilmos Zigmondy hit a gusher—hot water under high pressure. Soon the island became known as a therapeutic spa for those suffering from rheumatism, nervous disorders, and "female complaints." In the 1870s, the Grand Hotel was constructed, soon followed by the Dahlila, the Flora, and the Margarita, as well as elegant nightclubs and posh restaurants. All but the Grand, which was used by the Germans as a hospital and later by the Allies as a command post, were destroyed during World War II.

Kornfein remembers the twenties and thirties as Margaret Island's golden age. Still, enough elegance remains to lure him back. He winters in Palm Beach and spends his summers on Margaret. "But now I come for the cure, to be rejuvenated. You won't believe what the Thermal Spa can do to your body."

"Besides the pools we have various kinds of low- and high-frequency electric treatments—galvanic current, supersonic wave, interference current, ultrasound, microwave, and phototherapy," explains Dr. Hotchkiss as he strides through what seems to be an endless corridor lined with treatment rooms. "We use the high mineral content of the island's water to good advantage in our electric baths, during which the dissolved components diffuse through the patient's skin."

We pass rooms filled with lengths of rubber hose, chairs with leather straps and men in white coats attaching electrodes to squirming foreigners. I am confronted with what is either one of the world's greatest spas or a KGB interrogation center.

I wince in response to one of the civilized world's most disturbing sounds, the high-pitched whine of the dentist's drill. While I've always thought that a dentist is the very last person one would like to see while on holiday, a large number of West and East Europeans visit Margaret Island's dentists, known for both the quality and low price of their work. A filling costs about $9, anesthetic included. I wonder if some enterprising tour operator will cash in on this demand with a "Dentistry on the Danube" tour. Smile, you're on Margaret Island! Complimentary nitrous oxide! Certainly there are few islands where one can combine a holiday with some bridgework or have one's plaque or plate checked out.

Just when I'm afraid to open another door, Dr. Hotchkiss and Nurse Hamburger show me some more intimidating therapeutic machinery. We visit the Radiotherm 106, the Solar 1, and then Jules, who is much more menacing than any machine. Jules is a mad Hungarian masseur with a wrestler's torso, Popeye forearms, and hands the size of telephone books. A serious, tight-lipped fellow, he knows "Am I hurting you?" in eight languages, but ignores your response regardless of which language you select because he knows what's best for you. He reads a body like a CAT scan and always locates The Spot—that square inch of flesh where your tension resides. It is a joy to watch him work, though something less than that if you are on the business end of those incredible hands. Three-fourths of the earth's surface is covered by water and the other fourth by Jules's hands.

A worse torture than any machine can provide is inflicted on those patients required to diet while in Budapest. Dr. Hotchkiss gives each dieter a card, to be kept on one's person at all times, which lists the Dos and Don'ts. Dos are chamomile tea and lots of roughage. Don'ts are practically everything else. While as a rule no one with half a stomach left would undertake a tour of Eastern Europe for gastronomic purposes, Budapest is the exception to that rule. Hungarians *care* about food, as passionately as do the French. Eating is an event; fast food is a contradiction in terms. Few restaurants serve dinner without music. And no quantity of willpower provides sufficient protection against the temptations proffered by Budapest's cafés, pastry shops, cafeterias, and restaurants.

I didn't have the heart to tell Dr. Hotchkiss that yesterday in the city I observed two of his patients, a portly German couple, seated at the Vorosmarty, a famous pastry shop where the fashionable meet for coffee, tea, and sweets. The Germans look from their diet cards to the menu as they wrestle with their consciences. They secrete their diet cards back into pockets and purse. She orders strudel while he opts for *palacsinta,* Hungarian pancakes filled with sweetened curd cheese, chocolate, and poppy seeds.

No diet can hope to resist real Hungarian goulash, a very heavy stew, with tomatoes, green peppers, and pieces of meat, a meal in itself. Or *szekelygalyas,* a tasty combination of cabbage and sour cream, paprika, and pork. Or *racpunty,* deviled carp topped with potatoes, onions, and sour cream. Or *toltott paprika,* stuffed green pepper, and *toltott kaposzta,* stuffed cabbage. Or *paprikas csirke,* paprika chicken.

Sinfully heavy Hungarian cuisine is surely a diet doctor's nightmare. The health conscious can only console themselves with the findings of a Hungarian medical researcher who discovered that paprika is very high in vitamin C.

"Dyspepsia is the remorse of a guilty stomach." Who said that? Certainly not a Hungarian. A Hungarian may suffer an upset stomach, but guilt over a good meal? Never!

However, for those who do overindulge and have remorseful stomachs, the Thermal Spa offers a "drinking cure," which requires the patient to daily guzzle one, two, or three liters of Margaret Island mineral water. The water has a heavy sulfate content and trace elements from aluminum to zinc and is said to relieve gastric dyspepsia and assorted gall bladder maladies. Not only the dyspeptic imbibe; Margaritsziget Kristalyviz, as it is known, is the table water of choice in Budapest.

So many Hungarians swear by the stuff that I began wondering just what made it so good and so good for you. I decided to pay a visit to the state-owned Margaret Bottling Works, the one and only factory on the island. The plant bottles Hungarian soft drinks, Pepsi, and the island water.

I meet Lazlo Kuti, a tall, sober fellow who is chief engineer of the bottling plant. When I ask questions about the island's elixir, he grimaces and casts nervous glances at his even more sober assistant, the bespectacled and pretty Miss Erzebet Takacs, who takes copious notes of our conversation.

"Yes, the water is very good and we sell lots of it," Kuti says guardedly. "And please, we do not allow photo taking without government permission."

"When did the plant first begin bottling Margaret's water?" I inquire.

"Give me your passport, please," Kuti responds.

"How many bottles a year do you sell?"

"What is your mother's maiden name?"

"Is the water really good for the digestion?"

"Give me your press pass."

What we have here is a failure to communicate.

"All I want to know is what makes this water unique," I persevere. "What's the mineral content?"

Kuti recoils in horror. A kind of strangling sound comes from his throat. "You wish to know . . . the formula?" he asks incredulously in a hoarse whisper.

It soon becomes obvious that Kuti regards me as a spy, sent by decadent Western capitalists to his factory on a mission of industrial espionage. As Kuti shows me the door, Miss Takacs reaches for the phone.

During the next few hours my name is batted about Budapest. My suspicious activity is reported to all the Hungarian authorities—Passport Control, the Budapest Mayor's Office, party headquarters, and the Chemical Engineers Trade Union. All of these bodies soon determine that the mineral content of Margaret Island water is not one of Hungary's most important state secrets and suggest to Kuti that he was needlessly alarmed.

At the risk of creating an international incident, I must confess I smuggled the formula out of the country. Here's the secret of Margaritsziget Kristalyviz: $Na + NH_4 + Ca^2 + Mg^2 + Fe^2 + Mn^2 + HBO_2 + H_2S10_3 + CO_2 + O_2$.

It's printed on every bottle of Margaret mineral water.

Intrigue aside, Budapest's baths are much more memorable than its bottled water. A geological fault line occurs at the base of the Buda Hills, and the thermal springs, 40 million liters worth a day, bubble up from the ground and flow into Budapest's spas and swimming pools. One can follow this fault line and discover the history of Hungary's bath-loving conquerors. The Romans built baths in their province of Pannonia, which included territory in present-day Hungary and Yugoslavia. And in north Budapest, beside the ruins of amphitheaters and aqueducts, are found the remains of their steam baths. Lukacs Baths, built during the Middle Ages, now forms part of a rheumatology clinic. The Turks, who occupied Hungary in the 16th and 17th centuries, destroyed many fine buildings but built elaborate baths, some of which are still operating in Budapest. Here in this city, lying at the crossroads of Western and Eastern Europe, it's possible to find and indulge in most any kind of bath—including Russian, Baltic, Finnish, and Turkish. For Hungarians, the baths are a social center; they don't go in for private baths and prefer to stand around soaking and talking in a group.

Far more crucial to the well-being of the Hungarians than hot springs is another body of water—the Danube. For land-locked Hungary, the Danube has a powerful economic and cultural importance. Europe's second largest river (after the Volga) flows through eight countries on its

1,776-mile journey from its origin in the Black Forest to its mouth at the Black Sea. The river, blue in song only, is Budapest's crucial link with the west.

The Danube curves through Budapest from north to south, dividing repeatedly around islands. To the north, it slices through some mountains, makes a dogleg turn, and forms the elongated island of Szentendre, full of hamlets and farms. Hungary's new spirit of capitalism is visible on this island. The government provides small plots that citizens till after their shifts on state and cooperative farms. Profits are theirs to keep and high produce prices keep them happily gardening.

Just north of the city is the Isle of the Mosquitoes, and then comes the well-named Isle of the Shipyards. In the heart of Budapest is Margaret, and at the southern extremity of the capital stretches ugly Csepel Island, which features the mammoth Csepel Iron and Steel Works. Trucks, motorcycles, and bicycles are manufactured on the island, which is one of the most important centers of Hungarian heavy industry.

Other islands, not found on any map, are Hungary's new "Free Enterprise Islands." Certain sections of Budapest have been set aside for private entrepreneurs. State-owned shops and restaurants have been auctioned off or awarded to individuals who run them on a for-profit basis. About one Hungarian worker in ten is now employed in some form of private enterprise.

Budapest has become the economic showcase of the East Bloc. The stores are well supplied with consumer goods, and it is said that the only lines one sees in Budapest are at the cash registers. Austrians actually come from Vienna to shop in Budapest. The recently opened Hotel Victoria, about a block from Buda Castle, is the country's first privately owned hotel and the only one in the Soviet Bloc. A used car lot matches buyers and sellers by computer.

Hungary has stabilized its forint and now welcomes foreign trade. Even on tranquil Margaret Island I observed businessmen—German, American, Austrian, Hungarian, Swiss, and even a delegation from the People's Republic of China—walking along the Danube and making deals.

But Margaret is not a good place to make deals. Gazing down at you from across the Danube is Castle Hill, a vista of domes and spires, a thousand years of history, and it's easy to feel transitory, insignificant. Time moves very slowly here in old Europe, in this old city, on this old island. And time should be savored, as should this island's pleasures—a family reunion, a romantic stroll, a hot bath. [1986]

Leningrad: Soviet Islands of Culture

PAMELA SANDERS

"We shall meet again in Petersburg,
As though we had buried the sun in it" —OSIP MANDELSTAM

WE CAME IN JUNE TO THE city of Peter the Great, and Pushkin, and, yes, Lenin, traveling down from the north by train in the romantic hope of capturing something of what Peter might have seen in an expanse of Finnish marshes on the Baltic, and of perhaps sensing what Lenin might have felt as he returned in April 1917, smuggled by the Germans across Europe in a sealed train—"like a plague bacillus," said Churchill—to the Finland Station in what was then called, due to wartime sentiment, Petrograd.

"Remember, you're in the vortex of history," my husband used to admonish me good-naturedly whenever I complained about life in Moscow, which was daily during the two years we were posted there. As the train pulled away from Helsinki at precisely ten minutes past one o'clock on a Sunday afternoon, my heart began to beat faster and I recalled that old vertiginous spin of the vortex as I gazed at the accelerating landscape speeding past the window of the train that was carrying *us* to the Finland Station. I felt it again when, soon after passing Finnish customs at the border town of Vainikkala, we lurched to a stop in the middle of nowhere and heard, shortly, the heart-stopping tramp of heavy boots. An unsmiling Soviet soldier in olive drab slammed open the door and ordered us to stand while he lifted the bunk and searched the space beneath.

"What are they looking for?" my friend Caroline whispered.

"Stowaways."

"Do you think someone wants to be smuggled *into* the Soviet Union?" Indeed, it seemed a strange, almost touching conceit.

When they had finally gone, we ventured forward to the dining car over perilously heaving couplings and found the tables prettily set with white napery, shining silver and glassware and red carnations (plastic, but never mind, or as they say in Russian, *nichevo*) and a waiter who leapt to attention. Well, good on Gorbachev, I thought, and if this is *perestroika*, let's have more of it.

Taste buds quickening, we ordered open-face sandwiches of fresh caviar for two rubles and asked the waiter whether he preferred to be paid in Finnish marks or U.S. dollars. (The official exchange rate was $1.63 to the ruble, but the black market rate was four times higher.)

"Rubles," he replied.

"But," I stuttered in halting Russian, "how can I pay you in rubles when we're not allowed to bring rubles into the country?"

"How should I know?" he shrugged and snatched the menu from my hand.

SHORTLY BEFORE THE NEVA RIVER EMPTIES INTO THE GULF OF FINLAND ON THE Baltic Sea, it branches into three main arteries and countless smaller tributaries that coil and snake in typical river delta fashion around a cluster of small, flat islands. On these islands, now webbed and stenciled with streets and canals etched in granite and stitched together with wrought-iron bridges, there arose, like some Mediterranean fata morgana in the frigid mists, the strangest, most magnificent, most unnatural city that was ever devised by man. Peter the Great was the man, and St. Petersburg was the city. Named not after its founder, but after Christ's first apostle, the city became, within nine years of its inception, capital of the Russian Empire.

Perhaps the eponymous saint, a former fisherman and fisher-of-men, would have felt comfortable in these watery surroundings. With one significant exception, no Russian did. It was one of fate's little jokes that the only sea lover among millions of landlubbers in a landlocked nation should have been the tsar, a ruthless reformer and frustrated sailor without a sea who at the age of 12 had discovered a sailboat and, like many a yachtsman, never recovered from it.

When in the spring of 1703, while engaged in a protracted war with Sweden, Peter's forces broke through to the Baltic, the tsar decided to protect this newly won seaport with a fortress near the mouth of the Neva River. One of the smallest islands in the delta was the Isle of Hares (Zayachi), and it was here, on May 16, 1703, that Peter began to build his citadel, the Fortress of St. Peter and St. Paul, which was to be the nucleus of St. Petersburg.

The Russians say that Petersburg was "built on bones." Like most river deltas, the area was less land than bog; the very word *neva* (pronounced *neyeh*-va) means "swamp" in Finnish. Torrential rains, gale-force winds, and impenetrable fogs were constantly borne in from the Baltic; for half the year the islands were flooded, during the other half they were icebound. In these chill, miasmic swamps, between 50,000 and 100,000 workers perished from disease, starvation, exhaustion, and drowning. There was no housing, inadequate food, and only the most rudimentary tools. Sod was carried to the site bit by bit in the men's shirttails, wheelbarrows being unknown in Russia. On the adjacent Isle of Birches (now Petrogradskaya Storona), the storehouses, soldiers' barracks, houses of tradesmen and of the nobility, who came kicking and screaming only under direct orders from Peter himself, no sooner arose than they were washed away by floods or ravaged by fires, many of them set by the embittered populace. Wolves roamed the streets in packs, mauling and devouring hapless citizens.

Despite every setback, the city grew, and in 1712, a few years after his victory over the Swedes at Poltava, Peter made the decision which, to the intense dissatisfaction of absolutely everybody, made his beloved city the seaside capital of his earthly empire. After Peter's death, the fledgling capital was subjected to a royal tug-of-war, first abandoned by Peter's grandson, Peter II, who quickly moved the court back to Moscow, then reinstated by his niece, the Empress Anna, who just as quickly brought it back.

In time, however, the maiden city triumphed over the old "dowager in purple," as Pushkin called Moscow. Under a series of empresses—Anna, Elizabeth, and, most of all, Catherine the Great—and their relays of Italian architects, St. Petersburg, or, as its residents called it, "Peter" (the cozy diminutive was never pronounced "Pyotr" in the Russian way, but always "Piter," after the original Dutch spelling, "Sankt Pieterburkh"), truly became the capital of the aristocracy. For the next 150 years, this architectural gem of baroque and classical "ensembles" was mecca to dancers, artists, academics, and scientists, home to one of the greatest art collections in the world, birthplace of Russian literature—spawning Pushkin, Lermontov, Gogol, Dostoyevski, Gorky, Blok, Nabokov, Mandelstam, Akhmatova—and of the Russian Revolution.

The whole history of the Russian revolutionary movement is inextricably associated with this city—from the first strikes in 1749 and the abortive revolt of the Decembrists in 1825 to that day in November 1917 when the cruiser *Aurora* sailed up the Neva and fired the blank shot on the Winter Palace that signaled the start of the "ten days that shook the world." But within the year, Lenin closed Peter's "window on Europe," moving the capital back to Moscow. There is more than a little irony in the fact that after Lenin's death in 1924, the city he so hated should have been renamed in his honor.

BUILT IN 1970, THE HOTEL LENINGRAD GIVES OFF MULTIPLE EMANATIONS OF serious wear and tear: the crowbar marks on the elevator doors, the potholed concrete beneath the carpeting in the corridors, the slightly blackened patches around the electrical outlets. An urgent notice on the desk implored us in four languages not to use our electrical appliances—hair dryers, razors, and other newfangled exotica. On the other hand, the hotel had undeniable virtues. It had the merit, for one thing, of being in Leningrad. One hysterical tourist told me that she had arrived to find that she had no room and Intourist had resolved the problem by booking her into a hotel in Kiev. Our room was, moreover, perfectly clean, comfortable and cheerful, with two decent beds, sufficient if not capacious closet space, a functioning bathroom with—*mirabile dictu*—a bath stopper and plenty of hot water—albeit the rich dark brown of Russian sable, quite capable, from the look of it, of causing instant giardiasis or, at the very least, that other minor but disquieting malaise christened by some revisionist wit as the "trotskys."

Fortunately, there were alternatives to water. On every other floor of the hotel was a "buffet" (pronounced boo-*fyet*), where one could purchase bottled juice, sugar cakes, mineral water, *champanski* that came in two Zen-like categories—half sweet or half dry—plus a superior "brut" (we called it "brute"), and an alleged beer named Nevskoe Piva, which we decided must be drained from the Neva with a garden hose and lightly carbonated. We were grateful for these "bufyets"—as who is not who has suffered the assorted indignities of a Soviet restaurant?—but it was sometimes problematic to determine their hours or the currency they accepted. Ours took only rubles, of which, the moneychanger being closed on Sunday night, we still had none. Having learned our lesson on the train, we ordered first and asked later.

"*Nichevo*," said a handsome, strapping, jolly girl named Nina. "I shall buy them for you." We stood uncertainly, clutching our goblets of sizzling brute, nicely cooling down to tepid with the aid of some lumps of brown ice. "And by the way," said Nina, "have you any blue jeans, jewelry, perfume, T-shirts, stockings, cigarettes, chewing gum, or Reeboks?"

"Yes," said Caroline.

"Good," said Nina. "I shall come at once to your room and buy them all."

"But they're not for sale. They're ours," said Caroline.

"I shall give you lots of rubles."

"But we don't want lots of rubles. What would we do with them?"

That stopped her, but only momentarily. "Say, I love your jumpsuit," she said, fingering the fabric.

Caroline drew back with alarm. "I don't think it will fit you."

"No?" she eyed Caroline. "No, not for me. For my daughter of course . . . my little daughter."

Caroline's kind heart was already melting when I rescued her and, telling Nina we would meet her three nights hence to repay the rubles (she worked at the hotel only every third night, perhaps using the time in between to fence her acquisitions), we went to our room where we sat down in front of the window and raised a glass to the city that lay before us.

There are certain first sights that never leave one; the view of Leningrad from our window was one of them. Angled on the curve of the embankment, we faced southwest toward that sun that never seemed to set, looking upriver between the islands of the right bank—Zayachi, Petrogradsky, Vasilyevsky—and the unnamed concentric island-cities fanning out on the left. In the middle of our wide-angle view was the great bulging dome of Saint Isaac's Cathedral; on the left were the multicolored onion bulbs of the Church of the Resurrection, and on the right was the tall gold stiletto spire of the Cathedral of St. Peter and St. Paul, mirrored by the smaller Admiralty spire across the river. Except for these features, the city's profile was uniformly low, thanks to the tsars, who had decreed that no building could be higher than the Winter Palace, and to the Communist dictators, who for one reason or another had left it alone. There was not a high rise or skyscraper anywhere; they were all in the suburbs. Precisely because it was entirely horizontal, Leningrad had what few other cities have: a big sky.

Between the big river and the big sky, in the pearly 11 o'clock dusk, Leningrad lay stretched out long and low, massive yet delicate, an archipelago in stucco. Against the luminous "white night," we seemed to see the tracery of each tree in the Summer Gardens, each gleaming dome and fragile spire, each pilastered pastel facade, each scalloped bridge receding one after another into the mists while the glimmering auras of the clustered baroque street lamps and moving car lights and the iridescence of the sky all shimmered in the reflection of the river that cut like bands of cold colloidal steel through the great city.

Sometime after midnight, darkness fell, more like a diaphanous net than a curtain. By one o'clock, the net had begun to lift. My eyes closed, then opened. Something was moving, and I thought for a moment I was dreaming, but I was not, for all the bridges on the Neva were slowly rising and opening. For the next hour they remained thus, like huge dogs hunched along the river with their jaws open and eyes glowing, while beneath them in the violet half-light there came creeping a ghostly caravan of boats and ships and barges that stole silently as thieves out to sea. My eyes closed again and when next they opened it was six-thirty and the sun was shining and, looking out at this Italianate pastel mirage, I understood why people were always reaching for comparisons and striving for metaphors—the "Northern Palmyra," "Babylon of the Snows," and, most often, "Venice of the North." But really it was not at all like Venice, that ancient organic city thick with the naturally accrued grime and glory of a millennium of history, its waters swarming with life and vice and vermin. Here in Leningrad, there was no river or street life. It was all *tableau,* and no *vivant.*

Peter the Great had wanted St. Petersburg to be a water city like Amsterdam or Venice, but a major and finally unbeatable adversary in his grand design was that old foe, the weather. It was too wet, too wild, too unpredictable, and too cold for a waterborne populace. Much of the time the right, or north, bank was completely cut off from the rest of the nation and, before long, people began retreating increasingly to the left bank. But in the beginning, the working center of the city was on the Isle of Birches, for that was where Peter's cottage was, and wherever Peter was, so too was the heart of St. Petersburg—and, indeed, of the Russian Empire.

We went to see Peter's cottage on the Neva one crisp and breezy morning. Among all the monoliths and monuments, the mansions and the palaces, this log cottage is the smallest, simplest, oldest, and most cherished. Built in two days in May 1703, it is the only wooden house from that time still standing, thanks to the foresight of Catherine the Great, who took the precaution of surrounding it with a shell of brick. It measures only 12 meters by 5.3 and, given Peter's practical nature, it is surprising to what lengths he went to make it look like a little Dutch house; even the logs were planed and painted to simulate bricks. Inside the shell, the cottage is further protected by an iron railing put there so that visitors can lean upon that instead of on the window sills. Or so we thought.

"*Dyevochka*!" roared a fire-breathing babushka as I leaned upon the railing. "What do you think you're doing there?" Dyevochka means young girl, and I might have been flattered to hear myself addressed in this fashion had I not known that the babushkas, or old women, use this term to address any female under the age of 90. Indeed, we heard this rhetorical question put to us so many times during our stay in Russia that we began to think of it as the leitmotif of our visit. But the other side of this intrusively bossy village mentality is that the people are extremely warmhearted, and the same dragon who had seared my ears then went on, quite characteristically, to kindly show us around the exhibit.

These babushkas, who are the real enforcers of law and order in the Soviet Union, are also the backbone of what remains of the Russian Orthodox Church after 70 years of official atheism and harassment. In the pre-Bolshevik days when God was still in vogue, the Russian people were extraordinarily fervent worshipers. At the turn of the century, there were in St. Petersburg six cathedrals, 30 churches, and one of the largest monasteries in the world. In Leningrad today there are, we were told, five functioning churches. We found two.

On Saturday morning, Caroline and I attended the ten o'clock service at the Cathedral of St. Nicholas by the Sea. Designed by a pupil of Rastrelli named Tchevakinsky and built in 1758, it is a tall slender baroque structure, blue and white with five golden domes. Surrounded by a garden, it stands in a quiet corner of the city where the pretty, tree-shaded Kriukov Canal joins the Griboyedov. Outside the door, we were hailed by a stocky man in his 40s wearing a tan jacket with a gray shirt and brown tie. On his cheek was a mole like a raisin and on his head a bushy russet hairpiece with a low overhang like the eaves of a thatched roof, from under which he looked at us with bright blue eyes and an expression of shrewd good humor. He was the sexton, and we quickly accepted his offer to show us around.

There were, it appeared, two churches here: the downstairs for regular services and the upstairs for special services. Dimly lighted and pervaded with the scents of musk and incense, the lower church was typically Russian, with an iconostasis, numerous side chapels, a stone floor, and no seats. People came and went, mostly babushkas in black, lighting candles, some here and there kneeling, touching foreheads to the floor, repeatedly crossing themselves as they prayed, others kissing icons, their creased faces rapt with devotion and often wet with tears. Hearing some lovely plain chant, I walked up to the right side of the transept where a choir of four people was singing just in front of the iconostasis. Then I started visibly, for I saw over my right shoulder three old women laid out on raised catafalques. Side by side in their Sunday best, they smiled sweetly as they had probably rarely smiled in life, and though their eyes were closed in repose, they seemed to be slyly peeking at the friends who came to lay carnations and gladiolus on their breasts, peering out at them from under the freshly laundered kerchiefs which, even in death, they wore knotted beneath their chins.

Is it only elderly women who attend church now? At the morning service this seemed to be largely the case, but when we returned at six o'clock that evening for a service conducted by the Metropolitan, the highest-ranking prelate in Leningrad, in the upper church (far more splendid than the lower, it was brilliantly illuminated, with a towering iconostasis and the tsar's hulking black throne emblazoned on its baldachin with a golden double eagle) we saw many young and middle-aged men and women, some dressed up, others in simple clothes carrying string bags of vegetables or fruits, perhaps having come from a day in the country.

Between visits to the cathedral, Caroline and I walked up fabled Nevsky Prospekt, now as always Leningrad's principal avenue and Russia's most famous street. "There is nothing finer than Nevsky Prospekt," wrote Gogol. "What splendour does it lack, that fairest of our thoroughfares?" Wide as a river and straight as a blade, the Nevsky cuts a radial line through the city from the Admiralty to the Alexander Nevsky monastery, its two-and-a-half-mile length embellished here and there with imposing aristocratic fossils and religious incunabula—the strawberry-sorbet and custard-cream palaces of the Stroganovs, Anichkovs, Sheremetevs, Beloselsky-Belozerskys, and other great Russian families, now put to other uses by the Soviet bureaucracy, and the numerous deconsecrated churches, their crosses broken, their premises turned to pursuits less spiritual, even antithetical. The former Lutheran Church houses an indoor swimming pool (undaunted, the babushkas still cross themselves thrice while passing by), and the huge Kazan Cathedral, modeled after St. Peter's in Rome and the site of many a whopping tsarist funeral, is now the Museum of Religion and Atheism, with emphasis on the latter.

Those whose expectations of the Nevsky have been fueled by the vivid literary evocations of Tolstoy, Gogol, and Nabokov are bound to experience an initial sense of disappointment, however naive it may be—after all, the Revolution was hardly yesterday. Nevertheless, one finds oneself yearning for just a glimpse of its former elegance and cosmopolitan color—The Parisiana Cinema, The English Shop, Mellier's Bookshop, Abrikosov's Confectioners, Gratchev Gold and Silversmiths (the vaunted jeweler to the tsar, Fabergé, was not on the Nevsky but on Morskaya, now Ulitsa Gertsena, where the building still stands). But if the Gogolian "splendour" is gone, its celebrated air of hedonistic vivacity now replaced with a more stolid mood of making-do, there remains a sense of being at the center of the action and an atmosphere, even on this busy street, of friendliness. One has only to stand at a crosswalk or in front of a telephone booth looking helpless to be approached with an offer of assistance, a gesture that is all the more affecting given the trying vicissitudes of day-to-day life.

Starting at the handsome Anichkov Bridge of the horse tamers, we walked north on the great boulevard, crisscrossing from one side to the other, and what we saw, sadly, was pretty much the same everywhere—long queues of people waiting with a characteristically Russian combination of wry humor and cynical resignation to get into generally drab shops containing shoddy goods being sold for exorbitant prices. At meat counters, housewives picked up cuts of beef or chicken and sniffed them suspiciously; at record shops, customers took the discs out of their covers and with an obviously practiced eye cautiously examined them for flaws. In a pop record stall at Gostiny Dvor, the largest shopping arcade, a crowd of young people thronged around the counter trying to see what was available. One student said to his chum, "Well, what have we got this week—'Lenin's Greatest Hits?' 'Lenin and McCartney Together Again?' " In fact, the hot item was "Katz," a Swedish recording. But the other album being promoted was, to our amazement, not to say amusement, an ancient Pat Boone record. At the House of Books (Dom Knigi), there were only technical texts and political tracts, brown pamphlets with speeches by Fidel Castro and Kim Il Sung. The catchiest title we saw was *From Industrial Optimism to Technical Pessimism*. When I asked if they had any novels by Dostoyevsky, a salesgirl looked at me as though I had lost my wits. But the store was doing a brisk business selling huge posters of Lenin to American college kids to adorn their walls back at Vassar and Yale. In those prerevolutionary shops that still operated in their original capacities, the contents, needless to say, had changed drastically. The resplendent Eleseyev's, once the Fauchon of St. Petersburg, now the Gastronome (food store) #1, retains its art nouveau decor, but not its array of foods. Instead, one finds cabbages and onions and—that hallmark of Soviet display art—a pyramidal stack of tins with faded labels reading "Sauce" or "Jam." No confusing brand names here.

We found a touch of class on the Nevsky—the Literaturnaya Café. Once the Wolf and Berenger Confectionary Shop, it had been in Pushkin's day a literary hangout. It was here, in fact, that Pushkin stopped on that sad afternoon in January 1837 to pick up his second before his duel on the bank of the Black River with the French cavalry officer D'Anthès, whom Pushkin suspected of having an affair with his pretty, flirtatious wife, Natalia. Restored to a prerevolutionary ambience, the café has etchings of St. Petersburg on the walls, white silk Venetian shades on the windows, and ornamental trees of crystals and gilded leaves. For R24, plus a cover charge of R2 apiece, four of us lunched on consommé, caviar with cream, one crab salad, three fillets of beef, wine, ice cream, and coffee. As we ate, we listened to the strains of "Humoresque" and other light airs being sweetly played by a piano and violin duo. On our way out, the manager took us upstairs to show us an elegant dining room decorated in green velvet and featuring a grand piano where one can dine in the evening to classical music and poetry readings, a long-standing tradition. It was here that Lermontov read his poetic eulogy to Pushkin a few days after the great poet's death from the bullet wound suffered in his duel with D'Anthès.

Getting out of one's hotel for a meal was, we found, no easy matter. For the group tourist, it was doubly hard. The tour was so rigorously programed and the world outside one's hotel so fraught with complications that the visitor, already exhausted from a full day's schedule of palaces, had to be highly motivated to leave his hotel and venture forth on his own. But after one night of rubberized chicken *kotlyeti* and "La Paloma" played *con brio* by an eight-piece balalaika band in our cavernous hotel dining room, we were determined to be adventurous. There were basically four good things to eat, we concluded: caviar, fresh mushrooms, dark bread, and wonderful zesty soups like *akroshka* and *solyanka*. And one good drink: vodka. Such a meal was no hardship. Moreover, we found a couple of pleasant restaurants—the Sadko and, better yet, the fifth-floor restaurant of the Hotel Yevropeiskaya, which, with its *belle époque* decor and potted palms, was delightful. One night we dined less successfully—to be frank, it was a stupendous rip-off although rather fun anyway—at the once-smart, now shabby Metropol, where we were surrounded by riotous wedding parties all happily singing, dancing, and drinking themselves under the table. These were private weddings. Soviet state-run nuptials are something else entirely.

Weddings are "in," according to the directrice of the Central Wedding Palace, a splendiferous late-18th-century *palata* on the Neva that once belonged to Prince Andrei Romanov, a cousin of the tsar. In the German neo-Gothic library, a bridal party sat soberly awaiting its turn while listening to the brooding, piped-in strains of *The Godfather* theme. When the preceding couple had descended the grand staircase, the waiting couple ascended to a magnificent gold-and-white hall where the bride and groom—she in a long white gown, he in a lounge suit (obtainable at a bargain price, we were told, at the Jubilee Department Store)—were married by a stately woman in a long green dress, one of six "inspectresses" who officiate over the 30 marriage ceremonies held here each day but Monday. Attended by a small group of family and friends, the couple exchanged rings and kissed chastely to the swelling third movement of Rachmaninoff's Second Piano Concerto, followed by the Leningrad city anthem. Afterwards, in a vermilion sedan, the roof of which bore a large chrome ornament of entwined wedding rings, the couple repaired to a state-catered reception at a "Neva Dawn" banquet hall (one of six in the city), stopping en route to lay a bouquet of flowers at the base of the statue of Lenin atop an armored car.

Dining out is an almost certain way to meet Russians. One evening, we had a meal at the Troika, where we watched a floor show featuring eight gorgeous girls in strategic sequins and feather headdresses à la the Lido. Just before the show began, a man and a woman, rather to our surprise, sat down at our large ringside table. A newfound friend named Rodia had told us that restaurants were run and patronized by underworld figures and, to our amusement, the man was a gangland cliché in a pinstriped gangster suit, black shirt, and white silk tie, while his companion was a smashing baby-faced blonde. When the show ended, the *blondinka* began go talk to us in breathless Russian laced with charmingly broken English. Within moments, Andrei was buying rounds of champanski and we were all up and dancing. Andrei, who had been a dancer at the Kirov, was now, according to Svieta, the "owner" of a shoe factory where, she said, "he make many beautiful foots." He was, she said, immensely rich. Queried further as to what this might mean, she replied that he had a flat with *three* rooms. When we later asked our friend Rodia about this, he said that indeed, shoe factories had for some time been run by privately owned cooperatives—why, he did not know, but he had heard that government officials had had so many problems obtaining leather, etcetera, that they had finally given up.

Svieta, we learned, again to our surprise, lived in one room with her husband, a medal-winning gymnast, in a communal flat that they shared with three other families. Adroitly dumping Andrei and leaping nimbly into our car, she took us home to meet Sergei, "my beautiful husband," who proved to be handsome as advertised and wonderfully sweet as well. Besides teaching gymnastics, he was studying English. She was helping him. Together they proudly recited his most recent dialogue, the last two lines of which were unforgettable.

She: Are there some good French restaurants in this neighborhood?

He: Of course. It goes without saying.

But our most memorable night in Leningrad was spent at the Kirov Theatre watching, from a small bench backstage, the performance of this year's graduating class at the ballet school. They were 18, all vibrant with youth, beauty, and expectations, and committed to only one aim—to dance. To watch these lovely, lithe, leggy youngsters glowing with the thrill of their first performance before an audience was such a joy that it made one's throat ache. In the wings were a crowd of teachers, classmates, wardrobe mistresses, and makeup ladies, and as the young dancers ran offstage they were grabbed and warmly hugged, counseled, congratulated, or consoled. When the curtain came down at the end of the evening and the students went on to take their bows, everyone backstage went on as well, all of them applauding each other. This ballet "family" had been through eight years together, day in and day out, through childhood and puberty, measles and mumps, sore tendons and shin splints, trials and triumphs. Now the fledglings were on the edge of the nest, about to fly or flop.

The Choreographic Academy of Ballet, which in May 1988 will celebrate its 250th year, is located on classically elegant Rossi Street. On the day after the performance, we went over to the school, which had that slightly mournful air that all schools have at the end of term. Six of the best students who had danced the night before were taking a class. I would have thought they might have been granted a day off after their big night, but when we arrived they had been hard at work for three hours. A former prima ballerina named Zubkovskaya was the teacher, a striking woman with jet black hair in a chignon who was obviously a stern taskmaster. Then Konstantin Sergeyev, the artistic director, arrived and began putting them through their paces, correcting mistakes of the night before. We were in a large exercise room with a steeply raked floor, a barre on three sides, one mirrored wall, a grand piano, and otherwise bare except for a portrait over the mirror of Vaganova, the great dancer/teacher who had devoted her life to the school. When we left two hours later, they were still at it—the girls in black leotards and tutus, sweating and panting, their hair straggling down in damp wisps about their flushed faces, and looking as cross as could be—a far cry from the glamorous creatures we had seen the night before, yet somehow even more lovely, if possible, in this bare room where Pavlova and Nijinsky, Nureyev and Baryshnikov had each had their turn at the barre.

On the following day, we turned to another art form, hiking the Hermitage, the museum founded by Catherine the Great and today the repository of one of the largest and most superb art collections in the world. Logistically, alas, the Hermitage is not without problems: Admission must be arranged through Intourist; guidebooks and floor plans are not available on the premises; signs are only in Cyrillic; museum guards speak only Russian. Aesthetically, too, the museum has some major flaws: Too many paintings deserve to be retired to the basement; lack of lighting in combination with the often impenetrable gloom or glare necessitates a constant jockeying for position in order to see the paintings; tidal waves of tour groups led by aggressive guides with bodies like linebackers and voices that could shatter lead crystal, not to mention silence, sometimes makes this a less than pleasant experience. But nichevo, for there are among the 65,000 items displayed (out of 2,700,000) in some 400 rooms enough glorious Flemish, Dutch, Spanish, and Italian works of art, plus, specifically, 25 Rembrandt masterpieces and a staggering collection of French Impressionists, to exalt the soul and remain in the mind's eye forever. Nor is that all, for the ne plus ultra of the Hermitage may well be the hoard of prehistoric Scythian and Greek ornamental art that, together with some trinkets of the tsars, constitute what is known as "The Golden Treasury."

Stripped down to a claim check, we set off under the gaze of jittery guards and the tutelage of a well-informed and only slightly contemptuous guide with whom we spent the next hour and a half being mesmerized by the artistry of chunky soft gold stags and crouching panthers and miniaturized renderings in filigree of horse-drawn chariots and battling warriors on diadems, necklaces, and temple pendants by craftsmen whose aesthetic sense was so exquisite and whose workmanship so fine as to make the gewgaws of the tsars look like the rather vulgar playthings of overgrown children.

Emerging several hours later from the Winter Palace, which is connected to the Hermitage and included in the tour, we sank into a heap upon a low stone wall. It was a lovely afternoon. We sat for a while watching the passersby strolling along the sun-dappled quay beneath the linden trees—the women bare-legged and clad on this warm day in flowered nylon shifts, the men in drip-dry shirts and mouse-gray shoes, the little girls in sundresses, their hair pinned with enormous pastel net hairbows. A few meters away, an old man had set up an ancient doctor's scale and in the new spirit of free enterprise was doing a brisk business weighing people for five kopecks apiece. Across the river, on the strip of sand in front of the Peter and Paul Fortress, we could just make out the hordes of bodies—not a few in head-turning good shape, as we saw later—lying hip to haunch in the sun.

In the late afternoon, a queue of perhaps 200–300 persons stretched from the front door of the Hermitage down the block, the infinitely patient Russian people waiting hopefully to see the exhibit. Could they possibly get in by closing time at six o'clock? I wondered which poor souls would stand in line all day, only to be told as they came into the home stretch that the ticket office had just closed. What would they do? Bite their lip? Go quietly berserk? Try again the next day? I had a vision of them at the Pearly Gates, inching their way through eternity while party *aparatchiks* and privileged tourists queue-jumped happily ahead of them into heaven.

There was an option for these outcasts. They could leave the City of Peter and visit the City of Lenin.

There is a City of Lenin and there is a Leningrad. They are two different places. The City of Lenin consists of all those hallowed halls and sanctified flats where Lenin sat, spoke, or slept, plotzed or plotted after his return to Russia. Intourist does not take Western tourists there, because the tourists, wary of Marxist-Leninist propaganda and unwilling to desert the glamorous Romanovs for the arriviste Ulyanovs, do not want to go. But those who wish to feel that old spin in the vortex of history should give them a whirl. Leningrad, after all, is not all palaces.

Although the city was rechristened in 1924, Leningrad did not begin to come of age until the terrible 900-day siege by the Nazis during World War II. To see row upon row of mass graves at the Piskarovskoye Cemetery on the city outskirts is a sight that makes one weep. Leningrad rose from the ashes in a remarkably short time. And yet, in a way, it never revived, for the heart of the city is still St. Petersburg, that mournful, moribund metropolis of the Romanovs. Like a great Sleeping Beauty, she lies forever dreaming of her handsome princes and lost girlhood, while proletarian caretakers peddle her dreams, pandering to tourists nostalgic for a way of life they never knew.

Thus, at a "White Nights Festival" staged by Intourist, Communist youths, costumed as 18th-century courtiers, pranced and postured through the Summer Gardens, dancing an artful minuet in the gloaming amid the gaping statuary that Casanova once ridiculed to an apparently bemused Catherine the Great. Afterward, we walked back to our hotel along the Neva quay and over Liteiny Bridge. As we walked the quiet streets on this June night, I heard the whispered voices of ghostly lovers—of Anna and Vronsky, Peter and Catherine, Herman and Lisa, Nicholas and Alexandra, Tatiana and Onegin. But in the luminous twilight of a magical Midsummer Night in this captivating city of romance, I heard no real lovers laughing, nor saw any couples kissing on the quay. There was no one about—only a few fishermen hunched pensively over the embankment, and a large black dog.

"*Pasmatritye,*" said one of the fishermen to the dog. "Look."

The dog jumped up onto the granite wall and looked down into the river. I looked too, but in the still water saw only my reflection. [1988]

Asia
AND
Africa

Zanzibar: Out From Behind the Spice Curtain

BOB PAYNE

IN THE HEAT OF THE AFTERnoon, down narrow, shadowy streets where I tell myself what I smell is just a mixture of unfamiliar spices, I am trying to replace the contents of my lost luggage. The shops are all the same: dark, tiny, ground-level holes in the walls of stucco-covered buildings that could most kindly be described as not having weathered well. The shop fronts themselves are heavy wooden doors, some with intricately carved doorposts and lintels, some inlaid with cup-size brass studs that would discourage the press of any angry crowd. Plastered to one door is a poster warning against the dangers of AIDS. On another is a placard advertising a swashbuckling pirates-versus-the-British film—and the pirates seem to be the good guys.

Considering that the only bargain I've seen is a 100-pound sack of cloves, my shopping is not going too badly. I've found a pair of khaki shorts almost my size and a souvenir T-shirt that says, in Swahili, "Don't worry, be happy."

I've also been able to find—each in a different shop—shampoo from England, shaving cream from Kenya, toothpaste from Germany, and a toothbrush "Manufactured under the surveillance of China National Light." The shaving cream came wrapped in an English-language newspaper, three weeks old, from which I learned that a man in Torquay, England, was arrested and fined 80 pounds sterling for using threatening behavior toward an inflatable rubber doll.

What I haven't been able to find is a razor. And I don't think I'm going to find one in this shop either. The shopkeeper and I have gone through the ritual greeting: "*Jambo.*" (Hello.) "*Habari?*" (What news?) "*Mzuri.*" (The news is good.) I've scanned the meager stock in his glass case and on his dusty shelves, I've repeated "razor" at several decibel levels, and, with little hope of success, I am going through the motions of shaving.

But praise Allah! As I bring my thumb across my throat in what could be mistaken for a slicing motion, the shopkeeper breaks into a grin and rattles off something in Swahili that must mean, "I know exactly what you want." He ushers me out of the shop, slams shut the double wooden doors, grabs me by the hand, and sets off through the crooked streets at such a pace that I again trail sweat.

We go past black-garbed women who turn to stare, past grease-stained men squatting in the dirt repairing bicycles, past men hunched over ancient, clicking sewing machines, past a shop selling the skins of small brown animals, past Budda's Auto Parts, to . . . a barber shop. A lathered-up man is leaning back in a barber's chair. Four or five other men seem to be waiting their turn. To the disappointment of my escort, I make it clear that I don't want a shave, especially since one or two of the nicks in the barber's long-bladed razor look as if they could have been the result of a sword fight.

Then begins a rapid-fire, barbershop-quintet discussion, accompanied by much flailing of arms and eventually involving anyone who happens to be passing in the street, about what exactly it is that I do want. Finally, a consensus emerges, and I am whisked off again, to yet another shop and another shopkeeper.

"Razor?" I ask him.

"Gillette?" he responds.

And from that I conclude that Zanzibar, which like Katmandu and Timbuktu is one of the world's truly exotic-sounding places, still knows how to cope with visitors from far-off lands. A good thing, too, because very soon this 640-square-mile island, lying in the Indian Ocean within sight of the east coast of Africa, may be invaded by the kind of visitor who arrives in groups aboard wide-bodied jets for stays of seven days and six nights, tips and transfers included.

ZANZIBAR'S RULERS, WHO DREW WHAT HAS BEEN CALLED A SPICE CURTAIN OF isolation around the island when it joined with mainland Tanganyika in 1964 to become the United Republic of Tanzania, have now recognized that tourism—long lumped together with the evils of colonialism, imperialism, and capitalism—appears to be a necessary evil.

One result is that the Aga Khan, the wealthy, big-spending, religious leader of the world's 15 million Ismaili Mohammedans, has already cleared land for a 200-room resort hotel. Italian investors have started similar projects. And the airport runway is being extended to accommodate the inevitable big jets.

The adventure travelers have already arrived. You can find them here and there on the miles of white-sand beaches so little visited that it doesn't pay for muggers to stake them out, or in the lush green interior, breathing deeply beneath the short, compact trees that supply the world with a major portion of its cloves. And you can find them, usually lost and always claiming not to be, in the mazelike streets and passages of Stone Town, which is the oldest area of Zanzibar Town, the crowded, dusty city where the majority of the island's people live.

Most globetrotters stay in hotels and guest houses that charge from $8 to $12 a day, including electricity and water (on the days when there are electricity and water), and eat at food stalls in the Jamituri Gardens, where the local dishes are delicious and inexpensive (assuming one has already built up an immunity to the general run of intestinal disorders).

A few of the travelers, however, favor dining on locally caught lobster or prawns at the government-owned Bwawani Hotel, which, at the scandalously high rate of $49 a night for a single room, is easily the most expensive accommodation on the island.

Some of these modern-day adventurers spend their days peeking through the windows of the house where David Livingstone of "Dr. Livingstone, I presume?" once stayed, or in the huge, musty Anglican cathedral, staring thoughtfully at the altar built on the spot where the whipping post stood when Zanzibar was the largest slave market on Africa's east coast, or having their picture taken at the ruins of Maruhubi Palace, beside the pool where a sultan used to select the catch of the day from among his 99 wives.

The more nature-minded among the visitors tour the natural history section of the National Museum, looking in vain—among the birds of Zanzibar, the butterflies of Zanzibar, and the bats of Zanzibar—for the cats of Zanzibar made famous by Henry David Thoreau's pronouncement: "It is not worth the while to go round the world to count the cats in Zanzibar." (There is nothing special, by the way, about the cats in Zanzibar, other than that today, as perhaps in Thoreau's time, they are in numbers beyond counting.)

Late in the afternoon, when the heat begins to lose some of its intensity, you can join these young wanderers in chairs pulled up to the rail on the second-story veranda of the Africa House Hotel, which used to be the British Club. While they sip reasonably cold Tusker beers or Coca- Colas, and discuss such vital issues as where to stay inexpensively in Nairobi, you can see them looking to the west, beyond the soccer game on the grass at the edge of the seawall, beyond the ghosting, backlit dhows, beyond the faint black outline of the coast of mainland Tanzania, to where the sun edges toward the orange-topped clouds that will determine whether the day is to end with yet another perfect sunset.

"I've heard there's already been people here looking to buy their own holiday cottages," said a red-bandannaed Brit who claimed to be working his way around the world as a professional golf caddy.

"The next thing you know, you'll be able to buy souvenir T-shirts everywhere," said Celeste, a Californian who had come to Zanzibar because, like me, she considered its name alone reason enough to make the trip.

And Diane, an Australian whom I never saw in anything but a tie-dyed dress that looked as if she had been wearing it since 1968, voiced the one word guaranteed to cast despair into the heart of any traveler contemplating Third World paradise lost: "McDonald's."

To find anyone who thought that opening Zanzibar to more tourists was a good idea I had to go to the Zanzibaris themselves. And I had to go no farther than the four or five taxi drivers who spent most of every day parked in the shade of the casuarina trees at the entrance to the Bwawani Hotel.

"More tourists, more spice tours," said Mohammed, the taxi driver who gave me the standard half-day tour of the sultan's palace, the island's highest point (502 feet), and the red-earthed, spice-rich countryside for only four times the price I later learned was charged by the Zanzibar Tourist Corporation for the same tour.

Another driver, Ahmed, who described himself as a Zanzibari first and a Tanzanian second (a relatively common view the mainland government is not pleased to hear), said that Zanzibar, unlike mainland Tanzania, had been wealthy before unification. With help from the outside, it could become so again. "Tourism could bring us help," he said.

"But what if it brings people who want to make Zanzibar more like the places they come from?"

"We are Zanzibaris," he said. "Like a flag, we shift with the wind."

Ahmed obviously knew something of Zanzibar's past. In fact, the wind—the monsoon blowing steadily in one direction for half the year, then reversing itself and blowing just as steadily for the other half—has played a major role in shaping Zanzibar's climate and history. It brings the two rainy seasons that make Zanzibar look like the Garden of Eden. And for centuries it brought the slaving dhows that made Zanzibar a hell on earth.

The name Zanzibar comes from a Persian word meaning land of the black people. Beginning thousands of years before Christ, the Persians, and later the Arabs, their routes determined largely by the monsoon winds, traded with the natives in the land of the black people. In the process they developed a language, Swahili, based on a Bantu system of grammar but containing many words derived from Arabic. Today it is spoken in its purest form on Zanzibar. At least, that's what the Zanzibaris say.

The trade was in gold, ivory, spices, and—most profitably—slaves. Zanzibar, with its big harbor and its nearness to the mainland, developed into the center of the slave trade. On the mainland just opposite it, there was, and still is, a town called Bagamoyo. The name means "Here we leave our souls."

By the mid-1800s Zanzibar's slave market exported some 20,000 slaves a year, mostly to Arab countries. Its ruler, Seyyid Said, the sultan of Zanzibar, was so powerful that he controlled a million-square-mile area of Africa stretching all the way to the great inland lakes. As the saying went—in Zanzibar—"When the flute plays in Zanzibar, they dance on the lakes." Zanzibar acquired, and lived up to, a reputation as one of the evil places of the earth. Its dhow-choked harbor was a foul-smelling cesspool, its close-packed town was such a breeding ground for disease that on several occasions cholera took tens of thousands of lives. Yet there were those, the sultan chief among them, who considered Zanzibar a paradise.

Seyyid Said got his first look at the island in 1828. For nearly a hundred years Zanzibar had been ruled, mostly in absentia, by the Arabs of Oman, a sandbox of a place on the tip of the Arabian Peninsula. When Said, then the ruler of Oman, boarded a dhow to ride the monsoon to the southern reaches of his sultanate, it was a business trip. He had planned to show the flag and subdue a few troublesome locals. But he got one look at Zanzibar, greener than any place he'd ever seen, and fell in love with it. He immediately decamped from Oman and moved the seat of his government to the island.

Said was a good businessman, judging from the wealth brought to Zanzibar with the slave trade, and a good family man, judging from the fact that he had 70 wives and 112 children. He also had something of a green thumb. When a Frenchman gave the sultan some clove tree seedlings from the island of Mauritius, Said saw their potential: He ordered that for every coconut palm that died or was cut down, three clove trees be planted in its place. The cloves flourished, and Zanzibar was on its way to gaining its reputation as the world's major producer of the spice.

Seyyid Said made Zanzibar a force to be reckoned with. But what he did not reckon with was the force of the West.

In 1854, when Thoreau made his statement about Zanzibar felines (by which he meant, I think, that travel is not necessarily enlightening), Zanzibar was probably as far away and as exotic a place as he could imagine. For most Americans (and Europeans) of the time, Zanzibar was a dimly lit window into a dark continent so little known that just a few years earlier the European discoverer of nearby Mount Kilimanjaro had been labeled a fraud by the scientific community for claiming to have seen a snow-clad peak so close to the equator. The most famous of the African explorers—Burton and Speke, Stanley and Livingstone—all of whom would use Zanzibar as a staging area, were still a few years away from their historic journeys. Still, in 1841 the British opened a consulate on the island, and soon began the process of turning the sultans of Zanzibar into puppet rulers who would dance to the tune of European colonialism until the early 1960s.

In 1890, after some haggling with the Germans, who for a few decades controlled much of East Africa, Zanzibar became a British protectorate, although still nominally ruled by the sultan. Its wealth and power enormously diminished by the abolition of the slave trade, it remained under British administration until December 10, 1963, when it became an independent nation—for about four months.

On January 12, 1964, the African majority revolted against the sultan, and most members of the wealthy Arab and Indian minorities were given the ax or the boot. On April 23 Zanzibar and the neighboring island of Pemba joined with Tanganyika to become Tanzania. The young nation had hopes of becoming a model for self-reliant socialism in Africa. But by all accounts the model, based on collective farming techniques, never took off. "An economic basket case," is how one observer described Tanzania in a 1985 article in *Time* magazine. The problem was compounded by a decline in world demand for cloves, the major source of the nation's foreign exchange earnings.

Now the government seems to be looking for the parts, and the glue, with which to build a new economic model, based on tourism.

And tourism is already having an effect on Zanzibar. I had that demonstrated to me one day when four of us hired a van to ride across the southern end of the island to the beach at Bwejuu.

In the morning, in the dusty marketplace where a thousand Zanzibaris seemed to have some business connected with the loading and unloading of a half-dozen buses, we met at a prearranged spot. The van was already waiting, and waiting with it were the young driver, who was wearing a Bob Marley T-shirt, and Ali, the driver's interpreter, who spoke English about as well as you would expect of someone who had learned it in a schoolroom in a country that had been isolated from the West for more than two decades.

After reconfirming the price, which was not too much more than the price we had agreed upon the previous evening, our party climbed aboard and the van pulled out onto busy Creek Road. Before we were even fully onto the road, the van stalled and, despite the driver's grinding the ignition for two minutes, refused to start again. I remarked that we might be out of gas—a suggestion, I gathered from the driver's muttering, that was not particularly well-received.

"Plenty gas," said Ali.

The driver got out, made one slow perambulation of the van, got back in, ground the ignition a few more times, and muttered to himself again.

"What did he say?" asked Nigel, a tall, slender Christian missionary worker based in Mombasa, a Kenyan coastal resort he described as similar to Zanzibar, with the rough edges already smoothed off.

"He says you pay now, we buy gas," said Ali.

"No. If it doesn't start we don't pay anything," said Donna, a Chicagoan who had something to do with banking. Nigel, who had been in Africa longer than any of us, counseled patience.

The driver and Ali talked for a few minutes, glumly, it seemed to me. Then the driver appeared to be struck by a wonderful solution. He pulled out a crumpled wad of Tanzanian shillings and waved them in the direction of a place in the market that sold gas.

"In Africa," said Nigel quietly, "the answer is almost always patience."

Ali happily got out and pushed. We all happily got out and pushed. A group of boys in the street, engaged in one of the 70 or 80 soccer games that are taking place on Zanzibar at any given moment, happily abandoned their game long enough to push. And soon we were driving out of town on what the four of us were repeatedly assured, to the point of making us have our doubts, was a full tank of gas.

With reggae music blasting from the driver's tape player and red dust pouring in like surf through the windows that wouldn't roll up, we flew south at a suicidal 25 miles per hour on the cratered tarmac surface of a road that could not have had a nickel spent on its maintenance since the day the British left. The driver and Ali shouted greetings to people all along the way.

We slowed down only when a particularly jarring bump would make it necessary for Nigel to reach under the dashboard and reconnect the wires to the tape player. And we stopped only at the Jozani Forest, which is said to be the world's last red monkey sanctuary.

To help maintain the sanctuary, visitors are required to make a contribution equal to one American dollar, payable in hard currency only, unless, as was the case with us, the top man isn't there. Then you can pay in anything, because it is going no farther than his assistant's pocket.

When we reached the beach at Bwejuu we found that it was as lovely as we had been told it would be. Sixteen uninterrupted miles of white sand. Shallows for wading in. Sun for baking in. And enough coconut palms for everyone to sleep in the shade of his, or her, own.

I was looking for my own when I spotted a small boy sitting beside a pile of palm fronds that he was weaving into what I assumed was roofing material. He made such a perfect portrait of innocence and industry that I got out my camera.

I'd already learned that outside Zanzibar Town many people will run like frightened deer or become angry at the sight of a camera. But he grinned, so I took my picture. When I was done he said something to me that I asked Ali to translate.

"He wants you to show him the picture," Ali said.

I was amused, but embarrassed, too. The boy had apparently taken it for granted that I possessed magic powers, and, as painful as it was, I was going to have to admit otherwise.

In my explanation of why a picture would not be forthcoming I think I may even have used that very word "magic." Along with "no Fotomat."

Ali translated. The boy studied me for a moment, then spoke again.

"He says," Ali told me, "the last man could show him a picture right away."

Yes indeed, I concluded, tourism has already arrived on Zanzibar. [1991]

Seychelles: Do You Sincerely Want to Live in Paradise?

HERBERT GOLD

THE SUN DISAPPEARED. IT was red as sin, as the final good-bye, behind purple engines tumbling furiously across the sky. Then glasses crashed, people ran for cover, the hotel buttoned up. Storm: water crashed on granite and coral, waves plunged in grottoes and over rocks, the sea tore at the beach. I considered whether we would survive this onslaught; there was so little of us protruding above the Indian Ocean, our mountaintops dashed all around with sea. But the music on the radio assured me that all was normal: Simon and Garfunkel tapes about Mrs. Robinson.

You will note we survived.

During the downpour I ran on the beach. A small nut bounced off my head as I ran through the hotel grounds toward my room. Reality impinges even on American jogging.

As the evening steamed, the rain diminished, sputtering, and I studied the Coat of Arms of the Republic of Seychelles: a turtle supports a coco-de-mer. There are fish, sailing vessel, bird, sea. It's a sensible summary of what's happening, although the giant turtles are no longer eaten.

The people of the Seychelles do not choose the future (they have few choices about it) or the past (their history is quirky and ill-understood). They choose the present, they choose the life of granite and coral, sea and sky, on their precarious mountain peaks poking through the ceaseless ocean. These islands are events with precious astonishment for the rest of us.

Like paradise, the Seychelles are difficult of access. A few political undesirables—assassinating sultans, kings forsaken by their divine right, failed tyrants—have been sent to idle away their days here. Archbishop Makarios of Cyprus passed part of his exile in the house of Mr. and Mrs. A. W. Bentley-Buckle, a house named "Sans Souci." I had an introduction to the Bentley-Buckles from a lady in San Francisco whose former husband settled in the Seychelles. So there turn out to be faint connections in all this unreality with the somewhat less unreal world in which we who do not live on remote islands also pass our hours.

Various explorers and conquerors, such as Colonel Charles "Chinese" Gordon, have looked on this lush place with its erotic vegetables in X-rated shapes and its smiling R-rated climate and decided it was the biblical Garden of Eden. General de Gaulle, looking down his long nose at a possession lost by France, first to England and then to independence, in his great wisdom, piety, elegance, and snobbery called the Seychelles Islands "flyspecks." The few European residents and tourists sum up the ambience, geology, botany, trade winds, and white-sand beaches and name it paradise. A Californian, fleeing the hectic life of the American sixties, said he settled here because it was as far as he could go from San Francisco. He stayed because it was the Seychelles.

The archipelago, 1,000 miles from East Africa, houses a few more than 60,000 people, of mostly African descent, with curious and often beautiful mixtures of European and Asian adventurers bred into many of them. The languages are English, French, and Creole, that mixture of African and French which seems to have occurred almost spontaneously in Haiti, in the French Caribbean, in parishes near New Orleans, among peoples who have almost no contact with one another and yet finish by speaking the same language. Since I have lived in Haiti, and return often, I had a dreamlike sense of coming home to the Seychelles—of coming home to a dream and nightmare that I had never before visited. I had the same disturbing wave of feeling when I landed at Uli Airport in Biafra, during that terrible civil war, and heard starving Ibo children crying "agu, agu," and understood what I had always taken to be a French Creole: "Grand gout, grand gout"—*hungry*!

The people of the Seychelles, although poor and troubled by great power rivalries and uneasy politics, are far from destroyed Biafra. While oppressed by memories of colonialism and erratic self-government, they are not yet Haiti. Metaphors and comparisons might serve as warnings for the future; in the meantime, these hundred-odd islands, blazing white sand and blazing blue skies, granite sculptures and bays without a scrap of plastic or paper in sight, stand as blessing of providence scattered by the hand of some generous cataclysm of the earth. If God didn't will it so, perhaps, as one theory goes, they are the rock scraps left behind when India and Africa fell out. *Flyspecks*, Mon Général? Not for those who land and look.

Most of the people live on Mahé, an island of 55 square miles, where the major town, Victoria, contains perhaps 15,000 souls. A visitor who spends a few weeks on Mahé begins to recognize individuals, families, and to see the same people again and again. There is that family feeling of an island; we're in this together; it's manageable and we can learn our way around. We become instant native sons. The next step, of course, might be claustrophobia, island fever, wife-swapping, megalomania, alcoholism, reading good books, keeping a journal, and using florid language. I didn't stay long enough to experience each and every one of these delights—I have no wife, for example—but in two weeks I began to be able to say "I know your brother" to someone I had never met. And I was right. And after my first astonishment at frogs, beetles, lizards, creepies, and crawlies seen no place else on earth, as fruits shaped like sex and vegetables shaped like more sex, I realized that in fact the gregarious population includes much more than the people. The flora, fauna, geology, and history can keep a man company for years here. The Seychelles are a museum of a past we never had. Evolution has taken lonely and eccentric turns in this fertile spot to produce animals, birds, and plants that remind us of ancient history or perhaps the distant future. This is a laboratory for nature's experiments. The warmth and humid winds and isolation have conspired to construct a miraculous alternative to the real world which is still the real world, a paradise of the science-fiction unreal world, of giant snails and breathing grasses, of colors and transformations for the uneasily boggled mind.

The civilized history of the Seychelles is quite uncivil, as civilized history tends to be. First there was nobody, and that lasted a long time, except for a passing Portuguese or Arab or pirate, who touched soil, drank water, plucked fruit, and passed on. But in the eighteenth century the islands were finally settled by a few French and their slaves. By 1803 there were about 2,000 inhabitants of Mahé and the surrounding islets. They grew spice, pepper, nutmeg, cloves, cinnamon, a bit of sugar, a bit of this, and a bit of profitable that. Vanilla. Tea. They ate the jelly of the coconut. As to the fishing—well, it was almost as easy to fish as to breathe. They ate turtle meat and birds' eggs. Perhaps, like all Frenchmen, they also longed for Paris.

This quiet state was not to last; elsewhere wars claimed paradise, too. But here the extraordinary French governor, Chevalier Jean-Baptiste Quéau de Quinssi, deserves celebration as one of the great survivors. During the Napoleonic Wars, when the English warships sailed in, de

Quinssi rushed to meet them with a screed entitled "The First Capitulation of Seychelles."

When they sailed away he raised the French flag again. The process was repeated six times. Finally, in 1814, when the Treaty of Paris ceded the Seychelles to Great Britain, the game was over. So wise and prudent Chevalier de Quinssi became John de Quincy and ruled as the English commandant for 13 more years, until his death, which was the one capitulation he could not withdraw.

Friend Quincy's accommodation is an example of island agreeableness, palm bending to the storm, its roots firmly in place, its head nodding in the wind, firmly in the air. And just as he was both French and English, and then finally Seychellean, so the people who live here are African, European, Indian, Asian, efficiently stirred and mixed in the vat as in few other nations. If not colonial users or flee-the-world beachcombers, they are also Otherworldly.

Mahé is still a spice island. Victoria still boasts towers that resemble ones in London, only washed in hot water and shrunk down to a tight little miniature. Queen Victoria is only about a foot high. Paradise doesn't need tall queens and pedantic clocks. Just as the famous cathedral clock chimes twice—the first time to tell you they are about to chime the hour, then actually to chime it—so most of what one sees, smells, and hears one sees and smells and hears in a short while again. With all the steep green mountains suddenly dipping into white beach and sea, Mahé has retained the coziness of the Garden of Eden. No wonder Adam and Eve felt unhappy outside its gates, and no wonder Cain felt cheated.

An English former resident of the Seychelles said to me before I set forth on the long flight to London, then down the slope of Africa, across the Indian Ocean: "Oh, grief, woe. My house, my friends, my servants"—this is a summary, of course.

What he meant: History has found its way even here. Fertile emptiness brought colonists and slaves; the nineteenth century brought wars and more wars; the twentieth century brought an American satellite tracking station and a Soviet consulate stuffed with about 30 Marxist missionaries. After all, this is a nation with a population the size of a small suburb of Cleveland. But it is a member of the United Nations, it makes laws and sends out ambassadors, it is trying to join the larger world—it is trying to be a nation and to solve its own problems. Gilbert and Sullivan might collaborate with Eugene O'Neill to write the contemporary history of Eden; Alec Waugh did, in fact, write a book, *Where the Clocks Strike Twice,* full of fume, fun, and guerrilla Waughfare.

On the morning after the storm, the sun rose, in Joseph Conrad's words, "like a dazzling act of special creation," disconnected from the turmoil of the evening. These sunrises every day, with their unchanging equatorial brilliance, heralded by the loud waking of the birds, light up a land with few regrets, perhaps few hopes, but vociferous and boiling like the storm with desires.

ON THE WAY TO THE BEACH AT L'INTENDANCE, CAMILLE SUDDENLY STOPPED our Open Mini Moke—a kind of Australian jeeplet—at a twist in the road where he pointed to strange markings on the rocks. Part of the shore was dammed with pilings. There was a small depot of rusting machinery. "What is that?" he asked me.

"Is this a riddle? The Republic of the Seychelles atomic energy research laboratory?"

"No, an English," he said. "He had a treasure map. He dug. He spent his years digging. He died."

We drove on and talked of treasure. Great pirates, looting great ships, repaired to these islands for a bit of rest from raping and pillaging. They roasted turtles, drank coco milk, enjoyed their R & R. Eventually many of them were hung, but lacking numbered Swiss bank accounts some of them left deposits on Mahé. A few years ago a woman noticed odd markings on rocks on her land, dug, and discovered three skeletons with gold earrings in what used to be their ears. She and her friendly neighborhood notary public spent the rest of their lives digging but found nothing more. Other diggers have suddenly and unaccountably begun to live very well. No one is quite sure why.

At a barbecue on a beach one evening I encountered a crew of Australian oil drillers. I approached with caution. The non-natives were friendly. They explained what they were doing—digging dry holes for tax loss purposes. I understand a little Australian ("Stroin") and managed to converse with them. We performed the primitive beer bottle lifting ceremony, in which the head is raised and the Adam's apple wiggled in tribute to the Great God Lager. The leader of the crew explained: "There's bloody oil out here but we're not bloody lookin' for it. The boss company is a bloody cheatin' and murderin' multinational, mate."

"You might as well dig for treasure," I said.

"Same thing, mate. But tax write-off is the real pirate treasure these days, if you got a bloody brain in the bloody head."

Some of the most curious beasts to be found in faraway islands comprise the species Melancholic Runaway. They come in several varieties: the escapee from law or marriage, a comber of beaches, subsisting on alcohol and smoke; the remittance refugee, for whom life is difficult in ordinary civilization, and whose life is even more difficult for his family—

they pay him to stay away; and the Noble Unsavage, the well-off person who finds he can only keep up Standards in isolation, in beauty, in rupees plus cheap servants. The elegance of the last is not totally unrelated to the inelegance of the first. Often these foreigners don't like each other, but here they are, stuck in paradise together. Sometimes they meet at the Pirates Arms Cafe in Victoria, after a trip to the post office. They share the fish from the sea, the breadfruit from the trees, the toddy fermented from the palm.

In Haiti I knew voodoo devotees, Frenchmen and Americans who were convinced that the atom bomb, the Russians, or their former wives could not find them here, dissenters from Devil's Island, sexual specialists who required the cute company of very young boys or girls, aging collaborators with the Nazis. In the Seychelles I caught a whiff of these international subtypes in the bars of Victoria and along the roads where, for example, a European hippie in cutoff jeans blew a kiss at me as I rode past on my bicycle. Then he spoke in his mixture of French and Jive: "Tu me donnes quelque chose à fumer, man?"

Civilization is never totally out of reach, although frozen yogurt hasn't yet made it to this archipelago.

AT INDEPENDENCE, IN JUNE 1976, JAMES MANCHAM BECAME PRESIDENT, AND ruled also over starlets, a Rolls Royce, topless dancers, and European casinos, since Seychelles was too small to contain a man of his expansive poetic temperament. He seemed to see himself as P.R. man for Eden, more a greeter and maître d' than ruler, using the florid charm of an impresario to promote the tourist trade, often in pleasure palaces far from home. Like Walt Whitman, the poet laureate of another new nation, his spirit contained multitudes. He encouraged the making of soft-core porn films in the Seychelles—he was perhaps the only head of state to seek this trade. *Emmanuelle* was shot partly on the island of La Digue. In his grandiose and amused imagination, President Mancham saw the nation as a set provided by destiny for the benefit of X-rated impresarios.

His rival for power, France-Albert René, prime minister, looked on these shenanigans and decided they were not good. In the classical showdown, it was Mister Fun & Frolic versus Mr. Socialist Solemn. René's squad of revolutionaries smashed into an unguarded wooden outbuilding that contained the arsenal and declared Mancham overthrown. Since Mancham happened to be in London at the time, he simply stayed there. During "The Night of the Sixty Rifles" two people died, one drunk shot by a policeman, the other killed by one of his own allies. Shooting is not a Seychellean skill. The deposed president thought it foolish to break into the shed. "Twenty-five people with sticks," he announced, "can make revolution there." Only one of the martyrs of the revolution has a boulevard of Victoria named after him.

Socialism has brought regulation of drinking, prohibition of swearing, reduction of music on buses, and nervousness among foreign residents. Tanzanian "advisers" are training the new Seychellean army. I was denied entrance to a "Youth Village" by a Tanzanian in full parachutist drag. Surely the swollen Soviet embassy is not interested in helping to stamp out loud radios or keep the chickens off the Grand Avenue Francis-Rachel. Sex, a noncontrolled local enterprise like dominoes, produces little of value to the Russians. It may be that the way the Russians move in groups, huddled together for companionship, reflects fear of contamination by "the promiscuous islands." I saw five Soviet heroes negotiating the purchase of a bottle of Coca-Cola in the Victoria market. Then they climbed back into a black Mercedes 280 SL whose motor had been kept running by the driver so that the air conditioner would preserve the coolness of their Coca-Cola.

A few minutes later, behind the market, I found Magistrates Court in session and attended the trial of Marcel Moustache (alias Barbe Rose), accused of stealing an octopus from a fisherman. He claimed he was only buying it but ran off down the beach because the fisherman made a rude gesture to him. The judge was a slim English lad who looked as if he had been playing cricket at Eton two summers ago. The trial was conducted with the help of a Seychelloise who translated from Creole.

Verdict: Guilty. Fine: 150 rupees (about thirty dollars) or 15 days in jail.

Ironies: Here is a Marxist third-world country that still prefers to have its citizens judged by Englishmen. There's some color feeling, it was explained to me, and people are a little more comfortable with the boy judge from elsewhere—and besides, his salary is paid by an English aid program. We have here one of these radical Marxist third-world old-fashioned *colonial* type situations.

Partial moral: If you steal the octopus of another, watch out for the long tentacles of the law.

THAT NIGHT RADIO SEYCHELLES, THE GOVERNMENT-CONTROLLED AND ONLY broadcasting station, followed a denunciation of the United States ("the Conference of the Rich") and Zionism with the playing of Olivia Newton-John's performance of "A Little More Love." Radio Seychelles preached love, understanding, kindness, generosity, and the revolution of the oppressed. In other words the sun shines, the showers fall, the palms blow, the sea rolls toward the shore, the days pass, the birds sing. Marcel Moustache does his 15 days. A barefoot man with a straw hat strolls by my window, proudly carrying a 40-pound tuna. He is not necessarily proud because he caught it. He is proud because he has bought it—it's his. He is proud because he is proud.

Travelers have made much of the charm and accessibility of the female population. A longtime claim is that women outnumber men, and perhaps they do. Alec Waugh thought the girls "tender, caressing, ardent, gracious, natural, and utterly unvicious." They are shapely, well mannered, exotic, and friendly in ways that do credit to their French, English, Asian, and African heritages. The Paradise Flycatcher is one of the nice birds. The Lady of the Seychelles is another.

A popular belief is that rain on the day of a wedding indicates that the bride is not innocent: the rain represents the tears of the Virgin. Meteorologists, however, tell us that brief showers are usual most of the year. Q.E.D.

There is usually a worm in the fair apple of paradise, of course; this is the great tradition of Eden. The gonorrhea rate for Seychelles may be over 100 percent—every human being plus many of the trees. Visitors are advised to converse with the friendly people but to pursue only birds, fish, turtles, and wild fruit.

Considering these matters one afternoon at the Pirates Arms in Victoria, I was fighting off dehydration with the help of the Australian's favorite remedy. A man at the next table caught my eye. He needed a few words, a sociable meeting of the minds, to ease his way into evening. "Mmm, mmm, not bad," he said. He looked like a case of chronic liver trouble—eaten face, yellow highlights, feverish eyes, teeth not to be described. He sampled his drink again and remarked, "I find this is one's best comfort, don't you?"

He had been in the Seychelles since before independence. He made a living. He was liverish, liquorish, not very tall, and sad. As colonial exploiters go, he was more a survivor than a profiteer. But he had found a corner in Eden where he sold some product, represented some company, made out okay.

I noticed that the socks in his sandals were darned. In my country we either buy new socks or don't wear them. He was faithful to an older style, and I could assume his mother was not doing the darning. He had found a nice darner among the Seychelloises, and that was another reason for staying. He was a *coco-blanc*—a less grand blanc.

SINCE THE COUP THAT THREW OUT THE PLAYBOY PRESIDENT AND THREW IN THE Marxist Maximum Leader, there has been but one political party, the Seychelles People's Progressive Front (SPPF). Evidently the people are unanimous about everything, thanks to the perfect thought of the new guides to Progress. One effect is to make everything less complicated. It was my soothing habit at breakfast to read the morning *Nation*, published in mixed French and English by the Ministry of Education and Information; it is the only newspaper permitted. Some items are in French, others in English. One day the chief foreign story concerned "an agreement on cooperation in culture, education, and science between Socialist Ethiopia and the Mongolian People's Republic." The agreement envisaged "exchange of sport and art delegations." The local news was of winds collapsing roofs, delays in removing a boulder from a hillside, and confiscation of a poaching Japanese fishing vessel.

The lead editorial congratulated those who had voted in the party elections, "reaffirming the triumph of the Revolution," one of whose slogans is: "Only a Disciplined People Will Be Able to Achieve Progress."

On the back page a brief item reported that President Reagan had been shot.

Despite the regime, the revolution seems to affect the life of a visitor not a whit—perhaps less than a whit. Residents complain that the state-run agencies that try to control the fish trade and food marketing have cut down incentives, thereby cutting down food supplies, but there seemed to be lots of haggling, bargaining, and traditional African initiative in the various markets I visited. At the Victoria market I discovered another group of pudgy-faced Russians who seemed to be planning to buy a fish. Elsewhere I saw Chinese, also in groups, perhaps huddling together out of shyness.

Each village now has a headquarters of the party, the SPPF; on one near Victoria I saw painted a puzzling slogan: FEARLESS INDIGNITY.

There should have been a space after the first syllable of the second word. The IN was meant to be a preposition, not a prefix. It must have been lettered by a running-dog lackey of the bourgeois counterrevolution.

On the evening news Radio Seychelles announced two important advances. "The Specialist Room of the Hospital is now equipped to handle the most advanced cases of heart surgery. Soon we will bring patients from elsewhere in the world for cardiac treatment with all the advantages of our modern delicate equipment. We can also treat asthma and diarrhea in the Specialist Room with its modern equipment and its new face-lift."

At the same time the government announced a plan to produce salt from seawater so that the free progressive republic of Seychelles may be self-sufficient in this product. Salt and cardiac arrest, effective this evening and soon, respectively, are under control.

THE TWO-HOUR BOAT TRIP FROM THE MARINE CHARTER YACHT BASIN IN Victoria to Praslin and the Vallée de Mai is a voyage from never-never land to very seldom land. As we buck and chop through the waves and Mahé recedes, both Praslin and Mahé look like mountaintops covered with jungle, which is something like what they are. On the map I see a stony group of rocky outcroppings—islands? barely—called Les Parisiennes. They don't look like Parisiennes to me. Flying fish that look like flying fish, however, scamper over the waves. As we land, a seaman on board the motor launch assures me confidentially, "I know the word *chatouiller* in English—'teek-ell.' If you know the word *teek-ell,* it means you have made love in the language."

We debark into a small settlement nestled among the palms, a few houses, shops, bicycles, and a path leading into the forest preserve. Yes, I saw parrots and rare birds, fairy terns, widow birds, cheep-cheeps of all descriptions, also flying lizards and hanging bats, marvelous creatures for a creature-watcher. And I visited the lair of the coco-de-mer—that strange palm which was thought to grow at the bottom of the sea because its immense nuts, the largest in the kingdom of fruits and nuts, drifted ashore from the Indian Ocean before anyone had discovered palms here.

Why is the coco-de-mer so famous, so fascinating? You may have been asking this question all your born days and finally here is the answer. Its renown comes not simply because it is a rare and majestic tree, but because it flatters the human race as a sort of divine visual pun—a grand parody of human possibility. Put bluntly, it is a sexy joke on us by our Creator. So it seems. The female fruit looks like the lower shapely parts of a shapely woman, and few can gaze on it after the age of, say, 11 without either smiling or blushing. (The lady guide through the forest said, "And the jelly can be eaten. And the leaves to thatch your house. And the nut cannot be exported without a license—so many people come from everywhere to take our coco-de-mer . . . ") The fruit has been considered by many expert sheiks and sultans an aphrodisiac.

The male tree sprouts a warty, yellow-flowered phallic protuberance that droops every which way—most often gravely downward—to express Essence of World-weary Thick Tumescence. Surely it is anthropomorphism rank and shameless to say "to express" in the preceding sentence. I do not penetrate the intentions of the male coco-de-mer. It *seems* to express gratified desire in its lineaments, covered with little yellow flowers like something sold by Danish mail-order catalog.

Do the male and female trees dance together on the beach at night, as legend tells? Is that moaning and crackling the sound of their yearning?

No, alas. The sound is of their great leaves stirring in the wind, rustling like a potentate's fans. The branches creak.

Then how does the coco-de-mer fertilize and reproduce? I examined busy little bottle-green lizards nibbling at the yellow male flowers, shaking their pouchy necks, spilling pollen on themselves; and then they run to the female, and back and forth—the Pimp Gecko, I'd like to call it, in its showy bright green suit. Feeding bees, lizards, and sportive winds bring the two sexes together—not dancing in the Indian Ocean or even on the beach under the tropical moon. God's puns can only take us so far.

That night I imagined a coco-de-mer visited me in my natural habitat, politely inquiring, "And, sir? How do you fertilize, sir?"

And I answered, as I dozed to the sound of the waves on the beach, "I dance on the bottom of the Indian Ocean."

On the island of Praslin you have to like hot. You have to like humid. You have to like nature green and verdant in tooth and claw. If so, then you can stay the rest of eternity in this green spatter of the Seychelles, following the example of the tall grizzled old Scottish salt I met at the Flying Dutchman at Grand Anse. He has flat white slivers of beard on his cheeks, a mouth of distorted teeth, and hectic youthful eyes.

"When I first came, there were only two things to do," said the Scot, "spearfishing and fornicating. Now the former is illegal, and I'm too old for the latter."

"Then why do you stay?"

He grinned from on high. "I like it," he said.

Praslin, after Mahé, is the most populous island of the Seychelles. There are almost 4,500 inhabitants.

DO I MERIT SOME REWARD FOR MY VIRTUES? IF SO, OFFER ME A FEW WEEKS on La Digue, 20 minutes by boat from Praslin. This lower, flatter, seaswept island seems exactly the proper size for loneliness, gregariousness, privacy, and thorough withdrawal from the formerly real world (population: 1,911). On a sunny day I debarked, barefoot, in a place where the subway is replaced by the oxcart, the automobile by the bicycle. There are perhaps ten automobiles on the island—no rush-hour traffic, although my oxcart lurched in a burst of creaking speed when I whistled at a black parrot.

Sufferers from traffic anxiety and the proud owners of vintage sets of anomie might take a cure in this place. Just as God seems to have provided the coco-de-mer as a little play of nature on Praslin, here He offers us some modern sculpture: smooth granite clusters of rocks in the little inlet, sleek and unctuous, forming shapes as friendly as something commissioned from the hands of Henry Moore.

"The oxcart is late," said my guide, Rose-Marie.

"*Ne fret pas*, no hurry." I studied the bulletin board that announced a national song contest on the subject of Unity & Progress. A group of schoolgirls from St. Anne's parish chattered past in their blue-checkered uniform dresses, their pigtails, their smiles. The oxcart arrived.

It hauled me to lunch at a seaside house of straw, a roof over sand with kindly breezes blowing through. The serving persons were barefoot. The octopus was cooked in coco milk. The granite Henry Moores turned red, gray, red again, as the sunlight changed. The aspect of things has changed very little in the few hundred years of human presence.

A hardwood related to teak grows here (the local name for it is the musical word *takamaka*). It is used for boat building, for firewood. Palms, cinnamon, vanilla, and bananas scent the air. My oxcart lumbered through a plantation owned by a German millionaire. He has filled old fishing boats with shells; he has fenced the dirt paths. Giant land tortoises—some of them hundreds of years old—move like slow dusty dinosaurs in a rock-shaded pen. Their encrusted lurching is a visual equivalent of grumbling. The German provides access to the sea to oxcarts bearing visitors. Along the beach, open lodges with palm-leaf roofs house visitors from Mahé and foreigners from elsewhere in paradise gazing out at the sea or reading paperbacks of Simenon, Günter Grass, or supply-side economists.

I rented a bicycle and wobbled down the dirt roads when I tired of my ox. I looked at shells but did not gather them. I spotted another black parrot—two of them. I watched crabs dart back into their burrows at my coming. I studied the fish in the open market near the beach and felt the inexorable onset of dinner. I did not dread the prospect.

There is a stone church, a police station, an infirmary. There are village noises of hammers, saws, construction. Someone was building a fishing boat. They do it without blueprints—the plan exists in the head. I was offered a jar of homegrown patchouli, one whiff of which evoked Haight-Ashbury circa the Summer of Love.

Someone played a guitar. He sang in Creole. I could almost understand the words. He was not singing of Unity and Progress under the Progressive Front. He was singing of love, unity, and progress between man and woman.

Less than 2,000 people live on the few square miles of La Digue. The sea and beaches are impeccable. The isle is five kilometers long and three kilometers wide. It was discovered by Lazare Picault on June 17,

1744; the hour was that of sunset. Then, of course, there were no paved roads, and a similar situation prevails now.

WELL, VICTORIA (NEE PORT-ROYAL), DESPITE THE FAMOUS CLOCK TOWER, the famous miniature Victoria Station tower, the famous smallest monument to Queen Victoria, is no great capital on the order, say, of Dakar. It's an African village with a few extra rusted tin roofs, some fanciful gables, a market ("The Sir Selwyn Swelwyn-Clarke Market"), where you can find fish, fruit, batik, shells, and postcards. "Spitting Is Forbidden." Nearby stand the Barclay's Bank and a few Chinese and Indian shops (KRISHNA SUPPLIES). I was able to buy suntan lotion for an astonishing burn my normally resistant skin managed to achieve.

SPINIVA'S FANCYSTORE
Retailer of Goods
SHAM PENG TONG
General Merchant and Tobacco

A poster announced a Musical Show Celebrating 1981, Year of the Disabled, sponsored by the Victoria Baha'i Community.

The cathedral, on a slight rise, dominates the town, along with its ancillary stone buildings. There are a few banal modern structures; the hospital and the national library are graceful colonial houses, with ironwork balconies. There are the botanical and orchid gardens and a triste disco or two provide most of the formal entertainment. It is very hot. You won't find much of a boulevard for strolling, although the roads winding up the hills look inviting until you start to walk them: steep and hot. The Pirates Arms, with its café terrace open to the street, is the best spot for taking a cool Seychelles beer while watching the hot passersby watching you take a cool Seychelles beer. If you meet in town, it's the place where you meet, the Café de Flore of Victoria. My favorite sight turned out to be the shop called TRENDY BOUTIQUE with a Tanzanian soldier in full camouflage suit and paratrooper boots lounging trendily in front of a window display of Singapore Paris fashions. Since the current ruler, France-Albert René, prefers not to be overthrown as he overthrew his predecessor, James Mancham, he has invited the Tanzanians to stay for a while.

We don't openly discuss politics here, but I can look around. The National Bookstore carries the collected works of Lenin, Brezhnev, and other distinguished Soviet thinkers—a bit dusty, I noticed.

I browsed through a pamphlet transcript of a speech by President René in Cuba in 1979. After the usual denouncings and praisings, it concluded: "Long live the non-aligned movement and down with capitalism!"

The American embassy in Victoria is headed by a chargé d'affaires; the ambassador spends most of his time in Kenya. The Russians have a lot of people here. So do the Chinese. This dusty humid village lies athwart important passages in the Indian Ocean.

The busiest street in Victoria is known as *la rue du shopping*—a nice Creolization of two languages. In Gordon Square both Bastille Day and the Queen's birthday were celebrated until recently. The nation's own traditions are underdeveloped in the few years of independence and one minor coup. Cinnamon leaf oil is the essential ingredient in Coca-Cola, I was assured by a local booster who believes the Seychelles are central to American well-being. He puts his faith in cinnamon oil.

The sun went down. Evening falls with disconcerting suddenness five degrees from the equator. I strolled in cooling streets. The Flying Fox, that fruit-eating bat with a fleshy foxy face, was one of my startling companions. I can outrun the giant land tortoise. I was glad I was not a fruit when the foxbat dove above me with its sharp muzzle bared and its intentions unclear.

In Haiti there was a law that strictly stipulated that shoes were to be worn in the city of Port-au-Prince, but nowhere had the legislators specified *where* they were to be worn. I used to see the Madame Saras—the market women—undulating down the road from Petionville and Kenscoff with their shoes on their heads. In many ways the people of the Seychelles are cousins to this other Creole-speaking country so many oceans and continents away. The town limits of Victoria are defined by the place where the market women stop and put on their shoes. (Cousins differ, of course. Here they wear the shoes on their feet.)

The clock at the Cathedral of the Immaculate Conception chimed, then chimed again. Come on, hurry up, wake up, get to work or to your games. Ask not for whom the bell tolls twice. It tolls tolls for thee thee. It now tolled for me to take lunch at the Pirates Arms. I hope—sequence grows less serial here—I haven't already eaten lunch. If so, I'll surely only eat one dinner, and I'll run on the beach tomorrow at dawn, puzzling out once more how to add up the islands that form this nation.

There are 92. There are 89. There are 91. These conclusions come from research; I didn't count personally, and sources differ. There are over a hundred. Many are uninhabited by people. Most have birds, and many have crabs, critters, amphibians, turtles, dragon delights.

FOR NOT TOO MANY RUPEES I RENTED A GLASS-BOTTOMED BOAT TO TOUR THE rich coral reefs outside the Victoria Harbor. Mushroom coral, brain coral, yellow-striped coral, and marvelous parrot fish with bright parrot colors swimming in little seminars if not schools. My captain told me the really bright ones are the males, but if the lead male dies, one of the females becomes male, changes her colors, takes over new duties. In these waters we are Equal Opportunity Fish. There were fish that looked like pirates and striped fish that resembled zebras.

Then I put on my snorkeling mask and flippers and went into the water, taking care not to touch the sea urchins or the giant sea slugs. Later, back in the boat, a dark shadow passed beneath us—shark—and I asked the captain to lock up the snorkeling equipment.

We landed at Moyenne Island, which has had only four owners in the past 200 years. Brendon Grimshaw, the present proprietor—English, a former journalist, whimsical, a happy bachelor, gregarious—told pirate tales, turtle tales, ghost tales, and offered the visitors a beer. He is replanting the forest, preserving the ruins of the stone houses of the previous owners, and keeping an eye out for the pirate treasure everyone looks for.

People in these islands dream of treasure. Sometimes a stray soul suddenly becomes very rich and everyone whispers that he has found something. Brendon Grimshaw, the Master of Moyenne, living with his books, his slavery to the land and sea, his sunrises and sunsets, says he has found his treasure.

Less metaphysically, he has also opened a guest house on Moyenne.

Naked we come into the world, with oceanic yearnings, but the history of our future is inevitably a set of paradoxes. We grow more and more distant from salt, sea, warmth, and that sheltering closeness. We tell each other we are not alone; we are forever crowded by loneliness. Here in the Seychelles we can find our myths refreshed among the close and pressing facts. Coral accumulates from the sea, billions of creature shells piling up over the eons, and the world is reproducing and refreshing itself in miniature. The questing birds take note. What's this? A coral outcropping is a place to land for these constant feeders. They fly, they dip, they perch, they eat. When they perch, they do their bird things. Guano. Guano makes soil. The guano is rich in phosphate, because the birds feed on fish. Soil. The soil holds seeds that blow in the wind, or perhaps drop from the wings of the birds, or happen to be part of the birds' breakfast or part of the fish that was the birds' mid-morning snack. Trees. *Life.*

Accumulating from coral, new ones are growing.

There are a few more or less than anyone reports.

It may not be essential to count them. They rise from the sea, attracting birds, guano, soil, seeds, trees, and occasionally men and women.

The names of the smaller islands, some of them little more than coral or granite outcroppings, resonate with casual poetry: Silhouette, Frigate, Anonyme, Cerf, Thérèse, Félicité, Curieuse. Over 400,000 square miles of the Indian Ocean scatter and bathe this nation. Barely poking out of the sea, granite islands are probably the peaks of mountains left when Africa split away from Asia during the Ice Age. The coral islands may be as new as the granite ones are old. If the Garden of Eden was here, you can't prove it by me, but I can't disprove it either.

The Amirantes islands, another part of the Seychelles group, support vast bird populations. The Ile des Noeufs houses over two million pairs of terns. Over 100,000 giant land tortoises dwell on Aldabra turtle paradise, making the good life of a diet of birds' eggs. They are the same genus that can be found on the Galápagos. Some more numbers: over 880 species of fish, 120 varieties of shells, uncounted rare birds, zero panhandlers or muggers.

As a reminder for my golden years, I made a little list of some favorite odd tricks of nature in the Seychelles. (Nature might find some of our tricks rather odd too—the machine gun, Mace, PCB, woodpulp bread for reducing diets, indoor jogging tracks.)

The black parrot, which is gray and likes to mate.

The coco-de-mer, which is shaped like the southern part of a well-fed woman, only a little more submissive.

The flightless rail of Aldabra, which pecks fearlessly at visitors, perhaps looking for its extinct cousin, the dodo.

The black paradise flycatcher, known in Creole as the Veuve, or Widow, even the male ones.

The ghost crab. The brush warbler. The six-inch millipede, the husks of which kids pick up on the beach, break into crisp ring sizes, and wear on their fingers. Lizards, geckos, skinks. The moray eel—well, I could do without that. Frigate birds and various moaning, shrieking, and whistling creatures that, on the island of Cousin, keep a person awake as if he were in a busy canyon of mid-Manhattan. Fairy terns, shearwaters, marauding darters, and, alas, East African barn owls that were introduced in a foolish attempt to control the rats. (The owls ignore the rats and hunt fairy terns.)

Cousin is an island reserved to creatures small and smaller, and human beings are guests here, a meritorious and appropriate distinction. The litter of the birds provides rich soil. The litter from the people goes into dustbins. About this the Seychelles are strict—there is none of the mess of African compounds and great American cities.

The birth of land and land miracles, like mushrooms pushing from the sea, is continually in progress at Denis and Bird islands. A visitor can see this happening as if he stands in the amphitheater of creation. Truly such a privilege is treasure more amazing than pirate doubloons. Brendon Grimshaw is right. Everywhere the sea offers new beginnings, not just to the honeymooners, oil wildcatters, and stray good-timers of the beach hotels of Mahé.

Now, as I write this report in my flat on Russian Hill in San Francisco, a million waves and twelve time zones away, returned from an ultimate elsewhere, I crave to know more about the Seychelles—to visit Aldabra, where giant land turtles rule the island; to ask the mighty crab, which can open a coconut with one hefty pinch, to do its trick again; to ask the Tanzanian tribesman in the camouflage suit what he is thinking about as he stands guard with his automatic rifle in front of the Trendy Boutique in the capital village of Victoria. [1981]

A Taste for Penang: Malaysia's Marvelous, Movable Feast

CHARLES NICHOLL

GEORGE TOWN IS A CITY of snacks and aromas. All day long you walk around with your taste buds out on stalks while the corner cooks stir and fry, the warehouses waft out ginger and garlic, and vendors build baroque cones of shaved ice called *ais kacang*. Around teatime, down the wharf end of Chulia street, as the heat pours out of little printing shops producing the evening papers, you hear the welcome trill of the Jeplin Cake House van, which plays a Chinese song that sounds remarkably like "Daisy, Daisy, give me your answer, do." After dark, whole kitchen streets spring up, narrow and steamy and strip lit. At two in the morning you can still sit on a tin stool and dine on prawn pancakes and fresh starfruit juice.

It is estimated that there are 20,000 food hawkers in the city. Some of them push ingenious mobile kitchens with gleaming steel saucepans and complex gas piping. Others keep it simple, like the old woman sitting on a porch, attended by thin, hopeful cats, who stirs a pot of reddish broth over a charcoal fire.

This is the local noodle soup specialty, *laksa penang*. No two versions are quite the same, but the basic ingredients are fish, noodles, vegetables, chili, tamarind, turmeric, and a dollop of pungent shrimp paste.

"We call it laksa penang," she says, "because it has a little bit of everything."

AFTER A FEW DAYS ON PENANG, I UNDERSTAND JUST WHAT SHE MEANS. THIS is indeed an island with a little bit of everything, a tangy brew of different cultures and races simmering gently in the tropical sun. The basic indigenous ingredient is Malay, but the Malays themselves are now a minority of the island's half a million population, outnumbered by both immigrants (Chinese and Indians) and the annual influx of tourists (about two million last year).

It has all happened quite recently. For centuries Pulau Penang, the Island of the Betel Nut, was just another green dot in the blue Strait of Malacca. A few miles west of the Malay Peninsula, the island then was known to traders for its abundance of betel palms—source of the stimulant betel nut—and to Portuguese mariners who anchored in its northern bays on their passages from Goa to the spice islands of the East Indies.

Development began in earnest in 1786, when Capt. Francis Light took possession of the island on behalf of the East India Company. As an incentive to clear the land—so the story goes—he pointed the ship's guns at the jungle and blasted out several cannonades of silver coins. He later planted rice and pepper, and established the trading port of George Town, named after King George III. Onto the island flowed the Chinese—Hokkien traders, Hana tailors, Hainanese coffee shopkeepers, Cantonese goldsmiths—along with Indians and Tamils and Pakistanis, and the colonial British planters and administrators.

Every resident of Penang has a family story, that immigrant's sense of geography and destiny. Adrian Ooi is Hokkien Chinese. His father came to Penang and opened up a bicycle repair shop. In time he built this into a taxi business, then a bus company. Now Ooi is one of the leading travel agents in George Town. "Money is everywhere," he says from behind a perpetual smile. "It's just a matter of how you pursue it."

A couple of blocks away Rahim Karim sells batik cloth in a shop on Campbell street. His family came to Penang from Pakistan two generations ago. His grandfather opened the cloth shop that Rahim still runs, but his great-uncle moved on from Penang to Australia and became a camel driver on the famous "Outback Caravan" from Darwin to Adelaide.

Of course, the story is not always one of romance and success. O Chong is a ricksha driver in Penang. His father was also a ricksha driver. O Chong's son works at the Bata factory near the airport, turning out two-dollar beach sandals.

IN GEORGE TOWN, WHERE ABOUT FOUR OUT OF FIVE IN PENANG LIVE, IS THE best place to sample the island's melting-pot charm. It is a busy city, but the trishaw, or bicycle ricksha, is still the best way to get around. Most rides cost about 40 cents. At first I have some difficulties with the frank servility of the relationship—the lounging passenger and the sweating cyclist—but this is soon forgotten as we proceed through the streets, in eerie silence, at a pace tailor-made for seeing, hearing, and smelling the city.

I like a place where you can hear four different languages in a hundred yards. In Penang, conveniently, the bridge between them is English. (Anyway, the official language, Malay, is peppered with anglicisms, like *bas* for bus, and *ais* for ice, and *restoran* for restaurant.) Religions? Within a few minutes you can see the austere golden domes of the Kapitan Kling Mosque, the gaudy Hindu statuary of Sri Mariamman, the Anglican simplicity of St. George's Church and, most memorable of all, the marvelous Chinese *kongsi*, or clan house.

The kongsi is part meeting hall, part temple. The finest is the Khoo family's "Dragon Mountain Hall." Carvings encrust its three buildings: one for the family's patron saint, Tua Sai Yeah; one for the god of prosperity, Tua Peh Kong; and one for the "soul tablets" of the deceased, clustered like a miniature Manhattan of red and gold.

Adrian Ooi, the travel agent, was a Tao Buddhist. But four months ago he converted to Catholicism. He explains: "In Taoism, too many gods. A god for prosperity, a god for happiness, a god for this and that. For the Catholics just one god. It is more sensible." He smiles encouragingly, as if recommending a particular airline. In Penang people are proud of their cultural differences, but the mood that prevails is one of easygoing plurality.

The Chinese ingredient is perhaps the strongest cultural flavor in George Town. Indeed, some say that since the sanitizing of Singapore and Hong Kong, this is the most authentic Chinatown outside of mainland China.

But I also detected a strong, faintly musty recollection of English colonialism: a whiff of bygone tiffins, tea dances, and sundowners, a fossilized flavor of Penang in the thirties, when George Barnes was editor of the *Penang Gazette*, and *The Trail of the Lonesome Pine* was playing at the Rex, and the George Town traffic was a gentle succession of rickshas, bicycles, and bullock carts.

I feel I am back in that world the moment I enter the tall, airy vestibule of the Eastern & Oriental Hotel. The E & O is one of that select band of hotels, such as Raffles in Singapore and the Oriental in Bangkok, that are part of the white man's mythology of the East. More than any of the others—with the possible exception of the Strand in Rangoon—the E & O retains its atmosphere. In the Anchor Library bar, where I sometimes take an afternoon beer, I am surrounded by large-leafed aspidistras, mirrors advertising Player's Navy Cut tobacco, and faded sepia portraits of the Sikh Police Wrestling Team and the Chinese Literary Association. Deeper than this level of conscious preservation, there is a timeless somnolence to the place, as the ceiling fans gently stir, and the drinkers speak in whispers. Somewhere down a corridor I can hear the bellboy's shoes squeaking on the polished floor.

The era of British rule, which ended with Malaya's independence in 1957, is remembered with some affection by older residents. Haji Mohamad Mastan, born in Penang 73 years ago of Nagore Indian ancestry, presides over a rambling antique shop near the Kapitan Kling Mosque. In its murky depths, amid much junk, hide intricately lacquered dressers and large teak elephants. Overhead a 120-year-old crystal chandelier is draped in grimy plastic.

Haji Mohamad recalls the "British time" as one of order and decency. "The British were very strict; I appreciate that. The laws were all enforced. Now all is in a state of topsy-turviness." He lists various forms of topsy-turviness—people extending buildings without permission, Chinese setting off firecrackers at New Year, bicyclists riding without lights. I wonder if these aren't a small price to pay for independence, but the mustached Haji Mohamad is not a man to trifle with, and I say nothing.

"The people were *disk*-iplined in those days," he says, bringing his palm down flat on the cluttered paperwork on his desk. "There was no problem."

The latest of Penang's incomers are the tourists, more transient individually, but en masse having perhaps the most profound effect of all. Haji Mohamad remembers a different class of tourist: "They came here very well-dressed, in big ships. Now they come in very shabby dress, stay in cheap hotels. The young people bring in this *dadah* (drugs), and their dress, it is horrible." His hands trace a skimpy bikini line. "I think it is a very bad impression."

Probably dadah and bikinis pose less of a threat than sheer numbers. A third of the tourists who visited Penang last year were from mainland Malaysia and Singapore, with large contingents of British, Australian, and Japanese. Batu Ferringhi, or "Foreigner's Rock," once a north coast landmark for European spice galleons, is now virtually wall-to-wall hotels. At night the illuminated restaurants glitter like gaudy Christmas trees.

The increase in visitors has led to inevitable problems. Some say the sea around Batu Ferringhi is naturally cloudy, others say it is filthy with effluent: Probably both are right. I am told that a new sewage treatment plant is in place, and that once the big hotels are linked up to it, the water will be clear once more.

The authorities are aware that pollution is tarnishing the island's image. As the Malaysian prime minister, Dr. Mahathir Mohamad, recently put it: "Penang was once known as the 'Pearl of the Orient,' but now it is becoming known as the 'Rubbish-bin of the Orient.'" Unfortunately, much of the cleanup rhetoric is being directed at superficial problems like litter in the streets. Plans are afoot to ban food hawkers (denounced as "unhygienic") from the streets, and relocate them into "sheltered premises." In my view this would be a great pity.

The government did recently turn down a Japanese plan to develop a holiday village of hotels and condos, centered on a "mini-Disneyland" under a geodesic dome on Penang Hill, a haven of quiet and cool looming above George Town. But a $200 million project to create two artificial islands off the northern coast, linked by a bridge to Penang, may be in the works. This scheme carries the promise of a casino—a ticklish problem here. Casinos are very good business (especially with Singaporean tourists, who love to gamble), but gambling is not permitted under Islam, Malaysia's state religion. The idea of floating this profitable iniquity just offshore seems a convenient solution to the dilemma.

AWAY FROM GEORGE TOWN AND BATU FERRINGHI, THE PACE OF PENANG LIFE slows to a more timeless rhythm. A couple of bus rides (with reassuring notices saying "Do Not Spit" in four languages) take me down to Teluk Bahang, the last fishing village on the northern coast road. Chickens peck around dusty *kampung* yards, fishermen rest in net hammocks, and boys shin unbelievably up hundred-foot coconut palms.

Except for a high-rise hotel at one end of the village and three concrete housing blocks at the other, Teluk Bahang is a pleasant straggle of wooden houses along a bay redolent of fish in various stages of freshness. Many of the fishing families here are Indians from the Malabar Coast.

At a rambling guest house on the edge of the village I treat myself to the room called the "Tower Palace"—a former rubber smokehouse, with its own stairs and veranda, and tranquil views in every direction—all for the princely sum of 20 ringgit, about eight dollars, a night. I am told that in April, without even stirring from the veranda, I could pick ripe durian off a tree.

The owners are Cantonese. Mr. Loh is a retired policeman, former deputy superintendent for Penang State. I ask if there is more crime in Penang now. "No," he says, "it has always been here. Where there is inequality there is crime." A rotund, bespectacled man, he spends his days pottering genially around in his workshop, chain-smoking, and tap-tapping with his carpentry hammer. One morning he presents me with a new piece of furniture for the Tower Palace—an old radio cabinet, refashioned as a bedside cupboard.

Each morning I awake in a cockpit of sounds: roosters and dogs playing off the strange birdsongs. One day, walking west from Teluk Bahang toward the northwestern tip of the island, I at last glimpse something of Penang as it once was, before Light's silver coins cascaded down and the clearing began. The road peters out at the village jetty, but foot trails run on through lush forest, tonic and cooling, with strange peppery smells, and sudden bursts of animal movement: a monkey in the canopy, a snake in the leaf litter.

Once something disturbingly powerful and primitive crashes off through the undergrowth. An early warning device in my brain says Croc! even though I know it isn't. Instead it is a monitor lizard. (Later I see one swimming across a bay. From a distance it looks like a seal.)

The trail leads to a couple of beaches as lovely as any in Asia. Teluk Duyong, or Mermaid Bay, is long and palm fringed and classically tropical; Pantai Keratut, or Monkey Beach, is cupped and secretive, with a great green wall of forest behind. After a hot hour's hike to Monkey Beach, I am more than ready for a swim. I thought the place was empty, but as I walk into the surf, I see a man in the distance. He is running toward me, waving his arms.

It seems to take ages for him to arrive across the deceptive distance. When he gets near, he calls breathlessly, "No swim, no swim!"

He shows me his back. There is a great raw welt, the size of a tennis ball, bright red and ridged with white spots like a virulent nettle rash. It's from a jellyfish, the one great disadvantage of Monkey Beach. They are only there sometimes—when it's hot, I am told, unhelpfully. After a bit the man goes behind some rocks to relieve himself: One antidote to the sting—the only one available on a deserted beach an hour from the nearest road—is the ammonia in urine.

Meanwhile, I look northwest, toward Calcutta, across a thousand miles of open sea. Monkey Beach is the end of the line. There are no houses here, no cook stalls, no Taiwanese pop songs. But even here Penang is busy with life.

Rainbow-colored flycatchers dart to and fro, eagles circle in pairs above, and packs of rhesus monkeys move furtively at the edge of the trees. The monkeys, I'm told, can be aggressive, but I am glad they are here, glad there is still room for them in this overcrowded, all-embracing little island.

Like the jellyfish they are a few more ingredients in the Penang hot pot. Like each bowl of laksa penang, each day here tastes a little bit different. [1991]

Singapore: Crossing With the Green Man

BRYAN DI SALVATORE

Real Western Comfort in an Imperial Chinese Setting
SINGAPORE RESTAURANT AD

". . . He remembered muddy shores, a harbour without quays, the one solitary wooden pier. He remembered something more besides, like a subtle sparkle of the air that was not to be found in the atmosphere of today."
—JOSEPH CONRAD, *Almayer's Folly*, 1895

A FEW YEARS AGO, AFTER three months of surfing and hardscrabble wanderings along the thick, forbidding shores of some of Indonesia's more obscure and primitive islands, I landed in Singapore. I was malnourished, broke, exhausted, and in full retreat from the low-down, gone-wrong wilds of the Third World. My partner had already shipped himself back to the States with a near-fatal case of malaria, and I was soon to follow. Although my surfboard was still with me, I could no longer use it: I had a punctured eardrum, dozens of badly ulcerated reef lacerations, and a persistent and uncommonly vile strain of intestinal horrors. I caught sight of myself in a mirror in Singapore's Changi airport and was shocked; I was gaunt, my clothes hung on me like auditorium drapes, my sun-darkened skin was covered with large gray blotches, and my eyes were blazing and yellow.

Somehow I cleared customs, and was shuffling toward a bus stand when I stopped at a water fountain. I bent to drink, but then old warnings, water warnings, drifted across the channel of my soupy brain. I stopped myself.

I looked around then, for the first time. Changi was still under construction, but despite some confusion, it was as far from the last-day-on-earth chaos of other Asian airports as Juarez is from Wisconsin. Foremen and supervisors quietly discussed details with attentive workers; sign after sign begged my forgiveness for any inconveniences I was experiencing. Every inch was stone, glass, tile, or chrome. The place smelled not of bodies and ginger and stale cloves and rotting fruit, but like a cross between a doctor's office and the neck of a high school date.

This wasn't Asia, couldn't be. It was heaven or Oz or Zurich. It was an architect's rendering, some diorama of the future. Someone had played a cruel joke on a very ill surfer. I felt frayed, dowdy, evil, unclean. I began to cry.

I looked at the drinking fountain. I looked around me. Again the fountain, again around me. With a resolve, a faith that moments before I would scarcely have believed myself capable of, I bent and drank.

MOST ASIAN CITIES, ONCE ONE VENTURES BEYOND THE GLEAM OF THE CENTRAL multi-national interzone, are case lessons in the bizarre, the tragic, the inconvenient, the unequal, the unwholesome, the unhealthy, and the dangerous. But the city of Singapore, which occupies the southern fifth of the diamond-shaped island-republic of Singapore, stands amid these metro desperadoes like some well-scrubbed, well-fed honor student. The phones work in Singapore. The toilets flush. There are no pickpockets; few prostitutes. Instead of sprawling corrugated metal and packing-box slums, there are thickets of neat alabaster high rises.

The island is served by wide, multilaned freeways. People queue for buses and cabs. There are strict—and obeyed—seat belt laws, litter laws and jaywalking laws. "Cross with the green man," Singaporeans say earnestly, pointing to the light boxes at intersections. Singaporean skies are clear and blue, and one's collar is cleaner after an afternoon in Singapore than it will ever be in New York.

Shopkeepers do not block you on the sidewalk and tug you inside to inspect their wares. The only people who reach out in this city-state are polite young clerks, and they reach out for their calculators. How much? you ask. The clerk sighs, pulls out his black box, fans the keys and "Here, here is what I can do for you." Before your eyes, the tiny orange figures are tidy, irrefutable.

Since independence 21 years ago, Singapore has been on a dizzying spiral of upward mobility. While the concerns of its neighbors are crushing poverty, malnutrition and social instability, Singapore's officials issue somber edicts as to the correct placement of rubbish bins outside apartments, and express concern about the unsettling increase in the number of university students who insist—little thugs that they are—on wearing collarless shirts to lectures. Singapore's standard of living, by all generally accepted standards, surpasses most of South America and Africa, all of

Asia except Hong Kong and Japan, and even European countries such as Spain and Ireland. Its infant mortality rate is much lower than that of the United States, and Singapore, with one-third as many people as New York City, suffers fewer robberies in a day than New York does every 10 minutes.

But Singapore's success in shedding the more ominous of Third World burdens has brought it a steady chorus of criticism from both civil libertarians and romantics. The civil libertarians say that Lee Kuan Yew, Singapore's only leader since independence, has ruled with such an iron hand that material comfort has come at the expense of freedom of speech, spontaneity, creativity, and humanness. The romantics, however, despair of modern Singapore's reasonableness, its accessibility. They wish it was less Western, less scrutable.

Depending on whom you listen to, Singapore will either collapse any day, in some workaholic metro exhaustion, or it will close the every-day-tinier gap between it and Hong Kong and become Asia's premier entrepôt. Some see Singapore as a catalyst for Southeast Asia's economic gentrification, while others see it as an evil, moated parasite, draining its neighbors of what wealth and resources they presently possess or are capable of generating.

In any case, Singapore is rarely discussed in any but extreme terms. It has issued itself a mandate of perfection, and it has come so close to fulfilling that mandate that neither its leaders, its critics, nor its supporters can easily be pleased with less.

ALL THIS HOOPLA—THE DIRE PREDICTIONS, GRIM CURSES, AND UNRESTRAINED praise—is amazing for a place that 165 years ago was nothing but a malarial little swamp, home to a few fishermen, rats, centipedes, shorebirds, tigers, and low-rent pirates. (Even today, Singapore must import almost everything it needs, including water, and is a land of soil so poor that it is described, kindly, as "not even reasonably fertile.")

Originally called Temasek (Malay for Sea Town), the city was, legend has it, renamed by a fourteenth-century Sumatran prince who saw an animal "very swift and beautiful, its body red, its head jet black, its breast white." The prince declared the beast a lion, and Temasek became Singapura (Sanskrit for Lion City). For the record, the prince probably saw a tiger, one of many that inhabited the region. In 1857, for example, 300 people lost their lives to tigers in Singapore. The last tiger seen on the island was roaming a hotel lobby in 1932.

Once a minor outpost in the Sumatran Srivijaya empire, Singapore, under its new name, became a point of contention between two even more powerful empires, the Thai Ayuthia and the Javanese Majapahit.

The two forces bashed each other around a bit and sacked Singapore first from the north, then the south. Having destroyed the place, the two sides, in the best of imperial tradition, abandoned it. The Lion City became the Lion Village for the next 400 years, little more than a bush-league robber's roost.

Enter Sir Thomas Stamford Raffles. Dashing, handsome, ambitious, brilliant, a perfect candidate for a BBC miniseries, the former lieutenant governor of Java was worried sick that Britain, having recently returned Java to the Dutch, had not, in his words, "an inch of ground to stand upon between the Cape of Good Hope and China."

Defying official British policy, Raffles wrested control of Singapore and, in 1819, established a trading post there for the British East India Company. To do so, Raffles took advantage of a long successional dispute among local Malay royalty. The machinations of his operation were complex and, officially at least, the subject of British disapproval. Within three years, however, Singapore had become the most important company port in the region, and the British government was forced to, shall we say, abandon earlier moral objections.

Singapore grew from company settlement to colony, was occupied by the Japanese during World War II (the colony's guns faced south; the Japanese came from the north), and became internally self-governing in 1959. It joined with Malaya, Sarawak, and Sabah in 1963 to become part of the Federation of Malaysia, only to unjoin two years later, the result of a growing rift based on the fact that the majority of Singaporeans (75 percent of its 2.5 million people) are Chinese, not Malay. In 1965 Singapore became the United Nations's one hundred-seventeenth member and, as Lee Kuan Yew said, "the smallest of shrimps in a sea of hungry fish."

Raffles devised Singapore, rather than merely founded it. Nothing was left to chance. His policy of open immigration brought in thousands of badly needed laborers; his very forward-looking policy of laissez-faire capitalism—Singapore was, unlike other stations in the region, a duty-free port—all but guaranteed economic success. Raffles outlawed slavery, cockfighting, and gambling. He successfully discouraged speculation by declaring the best lands public. He laid out the streets, leveled the island's few hills, drained swamps, instituted religious freedom, and enforced the British judicial system. He rid the island of rats so successfully by offering a penny-per-head bounty that he followed with the successful elimination of centipedes. So lasting was his vision that at least one biographer says that the founding of Singapore was "the best thing Great Britain did for herself (or had done for her) in the entire century." The British East India Company showed its gratitude by hounding

Raffles, upon his return to England, for what they claimed was unaccounted for expense money. Raffles died in 1826, shortly after beginning the London Zoo.

WHILE SINGAPORE WAS WILDLY PROSPEROUS, THE SOCIAL FABRIC RAFFLES HAD envisioned was becoming a bit unraveled by the early twentieth century. The city was a brawling place, thick with sailors and laborers from six continents. Ruled by ruthless Chinese gangsters, it was a lawless, depraved, and dangerous place.

Even as recently as 1959, the country was wracked by unemployment, labor strife, and political logjams. The island was home to some of Asia's worst slums. Singapore, going downhill fast, was, in the terms of one long-time observer, "a goddamned mess."

But Lee Kuan Yew, through a combination of financial astuteness, a brilliant recognition of the workings of power and persuasion, and a nearly evangelical devotion to the work ethic, has brought his speck of a land ever closer to the economic, social, and moral ideals that neighboring countries can only dream of.

Certainly even the most skeptical observer will agree that Singapore has achieved an economic miracle. Singapore is the world's second busiest port, with a ship entering or leaving its harbor every 15 minutes. It is the world's third largest oil-refining center. Singapore's currency is stable, inflation is not a problem, unemployment is less than 4 percent, and many businesses are chronically understaffed. (At any given time there are at least 200,000 "guest workers" in Singapore, many of them from Indonesia and Malaysia.) Strikes are unheard of, worker productivity is nearly double that of Japan, and despite its dearth of natural resources, Singapore's balance of payment surplus in 1982 was just over $1 billion. Ninety percent of all Singaporean families own a refrigerator, a telephone, and a television.

One of Lee's most effective ways of bringing this economic and social turnaround has been a series of widespread self-improvement campaigns. There was a successful mosquito eradication campaign, a productivity campaign and a courtesy campaign, as well as campaigns against long hair, litter, and jaywalking. The "Speak Mandarin" campaign is now in full swing, as evidenced by the incessant and thoroughly maddening vocabulary lessons blaring from shopping center loudspeakers.

Over the years, another campaign, one promoting a form of twentieth-century Confucianism, has not only helped the other campaigns succeed, but has given a philosophical underpinning to Lee's vision of Singapore. Confucianism, briefly, demands of an individual, or a populace, absolute obedience to authority. This authority, in turn, will provide for its subjects' every need, as benevolently, generously, ethically, and fairly as possible.

In the presence of the dirty old real world, however, which is not perfectly harmonious, obeying the authorities can lead to some grousing and, very possibly, a passivity, a docility, a resignation, even a stifling of imagination, creativity, and dissent. "How many Singaporeans does it take to screw in a light bulb?" the current joke asks. "I'm sorry, but I'm not authorized to comment on that."

Little by little the hair of Singaporean youth creeps back down over its collar, even after a decade of anti-long-hair campaigning. Jeans are no longer taboo, and most people can get their hands on a *Playboy*, ostensibly banned from the country. Last fall, headlines in the *Straits Times* blared: "Police Pincer Traps 700." The story that unfolded concerned a roundup of motorcycle-riding youth who congregate each weekend on the East Coast Parkway, flirting, smoking cigarettes, and wearing shirts emblazoned with "fancy names like Jaguar, Hunter, USA, Jaws and Lumberjack."

In his most controversial campaign, Lee publicly bemoaned the fact that many of Singapore's university-educated women were remaining single and childless. Women without a university education, Lee went on, were reproducing twice as quickly as were graduates. Implying that nature easily overrides nurture and painting glorious pictures of a land that could be bathing in a crystalline gene pool, Lee predicted darkly that unless university women reproduce more often, Singapore's "levels of competence will decline, the economy will falter, the administration will suffer, and the society will decline."

IT MAY BE TEMPTING TO SNICKER AT SUCH WIDESPREAD CONCERN OVER WHAT, to a Westerner, might seem such mild antisocial flutterings. And it can be mildly embarrassing to witness such total acceptance by Singaporeans of unabashed paternalism. But while we might not expect Kalahari bushmen or Laccadive islanders to behave as we might in a given situation, it is oddly confusing, disconcerting when Singaporeans do not. And the reason for this confusion might be that, given Singapore's glittering Western front, we begin to believe it is, in fact, a Western place, peopled by Westerners. It is not.

For every McDonald's or Colonel Sanders, there are hundreds of street stalls and food vendors, serving up some of the most miraculously tasty Szechuan, Cantonese, Dravidian, Malay, Indonesian, Punjabi, Nonya, or Hokkien food on this planet. For every formica-rich hotel coffee shop, one might find a café next door with a wild-armed cook in a singlet frying something in a huge wok, standing behind stacks of blue, red, and yellow plastic plates and cups, pyramids of eggs, mounds of noodles, whole sides of savage red meat, bottles of mysterious sauces, and bales of green vegetables.

For every clinical, brand-new vertical mall, there are 20 Indian street-side drug and magazine shops, thick with Indian movie magazines and cigarette packs, hard candies, packets of white melon seeds, candy corn, headache powder, the ubiquitous Tiger Balm salve, cough drops, candy bars, paper tablets, stamps, toothpaste, matches, and pens.

For every tastefully appointed Gucci-Dior-Givenchy-Klein boutique, there is a dark hole-in-the-wall selling coal or wood or live birds or sari material or Hindu festival cards—Happy Deepavali to you and yours—and spangles and incense sticks.

For every international bank, there are a dozen private money changers, beating the bank's rates, no questions asked, cash only. For every shop selling Sony or Panasonic, there is a Fortune Tailor, a Lucky Awning shop, a Dock Han Ming Bridal Dress store, a Mohamed Thamby Limited, an Angel Garment Company and, of course, a Goh See Seng, "Sole Agent for Miscellaneous Products."

While there are Mercedes by the dozen, there are also rickshas, their drivers, with legs like steel bands, on the far side of 60. You can see them at night, asleep in their rigs, with a bit of wash—a T-shirt, a pair of shorts perhaps—hanging folded over a line stretched from the ricksha to the door handle of a 40-story building.

THE MALAY *KAMPONG* (VILLAGE) OF TANJONG IRAU FACES THE JOHORE STRAIT, which separates Singapore from the rest of Asia and is 20 miles north of metropolitan Singapore. It is a mile or so from the nearest paved road, at the end of a narrow, deeply rutted track.

The village is a loose cluster of a dozen raised houses, glassless windows open to the sleepy afternoon air. Just past the village is the beach. Small sampans are in the water, as well as a few yachts and large cabin cruisers from a nearby club. Chickens and roosters and ducks pick along the wet, dimpled sand. Higher, where the sand is dry, men in sarongs talk, sitting on upturned boats. Cats, children, and coconut husks are everywhere. The smell is low tide and wood smoke.

In the village square is a dirt court where the local swains, watched by their younger brothers and neighbors and village girls, are playing *sepak takraw*, a combination of volleyball and soccer. The ball, a rattan weave, is the size of a grapefruit. The boys are showing off, volleying with amazing over-the-head cartwheel spikes. The girls pretend not to be impressed as they pinch and elbow each other and giggle.

Two old men are asleep on a porch. Two old women chat out their windows, over the soft pop music coming from a large portable radio.

Like many other pockets of old Singapore, Tanjong Irau will not remain isolated or quiet for long. Across the track, a couple of hundred yards from the square, the palm trees and sea almond trees and thick, low shrubbery are gone, and the earth is red, bare and bladed, baking under the equatorial sun. As soon as the ground has settled sufficiently and enough workmen can be spared from other projects, one more of Singapore's enormous public housing estates will rise.

In Singapore, a new housing unit is begun every 17 minutes. Since 1960, Singapore has reclaimed an area larger than Hong Kong Island from the sea. Beach Road is now a quarter of a mile inland. Groves of pile drivers are everywhere; their incessant clanging has overpowered even the blaring Chinese funeral trucks.

Construction crews work around the clock, the huge gantries swinging atop building after building like monstrous metallic crickets. Dump trucks, detours, and Danger! Keep Out! signs are part of everyday life.

While construction projects in downtown Singapore are mostly private developments, billion-dollar world convention centers and office-hotel-recreation-financial complexes, the bulk of Singapore's island-wide construction is devoted to public housing, where over 70 percent of Singapore's citizens now live. These massive 10- to 20-story buildings are thick on the island, ringing the inner city like an alabaster reef. Each development, racially integrated to reflect Singapore's ethnic background, is an entire village, complete with shopping centers, theaters, places of worship, and recreational facilities.

Last year Singapore's relocation process entered Chinatown, perhaps the most famous of Raffles's ethnically segregated neighborhoods. The bulldozers roared down the streets of the two-mile cluster of shops, houses, street stalls, and sidewalk markets. The lantern makers, letter writers, statue carvers, clog makers, junk dealers, tinsmiths, and seal carvers will be relocated. They'll leave their dark shops and tin and wood street stalls and move to gleaming chrome and aluminum and glass, brightly lit high-rise malls a few blocks and a thousand years away.

No small amount of concern and grumbling and nostalgia greeted the decision to raze Chinatown. Even the *Straits Times*, usually given to echoing government policy, ran features on the demise of "Old Chinatown," complete with photos of sad, stunned, defeated octogenarian merchants watching their shops being shredded.

The villagers of Tanjong Irau, however, don't seem terribly concerned that some of them will be living miles from their neighbors of today, scattered around the island. Certainly they will miss each other and the quiet

Saturday afternoon sepak takraw games, but there is no evident outrage, no apparent sense of loss.

"When we get to an estate, we won't have to worry about insects any more, or dirt, or floods every year," one villager said. "We'll be closer to our jobs. We only fish for a hobby now." He stopped and laughed at a friend's remark. "If you live in Singapore, you get used to change. Besides, we won't be moving for a long time."

How long is that?

"Two, three years."

Two or three years can be ages in Singapore. In that time the bumboats can disappear entirely from the Singapore River, relocated by decree for sanitary reasons. In two or three years, all the seagoing traffic—the dumpy intercoastal traders, the tugs, the barges, the oilers, and the gargantuan container ships and tankers—can be moved so far out in the sea roads, so entirely blocked from view by reclamation projects, that on a cool afternoon or evening, Singapore can seem like anything but a harbor city.

In two or three years, the number of rickshas can decline dramatically, as can the number of guards asleep on their cots in doorways. In two or three years, a favorite corner Szechuan café can disappear under a wrecker's ball, along with a small, cool, dim Chinese hotel where you once slept for five days straight, recovering finally, strong enough to return home.

In two or three years, the Changi airport can be completed, and the airport can become a showcase, with waterfalls and excellent restaurants and huge aquariums embedded in walls and marvelous fern jungles between the tracks of softly whirring luggage bays.

And with a little luck, a fellow can return to Singapore and imagine that he has found the same drinking fountain that so astounded him two, three years earlier—an entirely unremarkable drinking fountain, really, but one that tipped him off that somehow Singapore was not like any other place in Asia.

He can watch Singapore continue on its hectic way, and declare it good. With the generosity so often born only of a full stomach and bank account, that fellow can shrug off troubling questions about Singapore's soul or lack of it. It is more important, he thinks, remembering his own days of raw-bone travel, that no one is starving or is likely to for quite a while. He can look around and realize that while Singapore isn't—quite honestly—heaven or Oz, it might just get close to it someday.

But it is amazing how things change, that fellow will think, in just two or three years, as he catches sight of himself, or almost all of himself, in an airport mirror. [1984]

Because of You, Kyushu

RITA ARIYOSHI

THERE IS A SAYING IN OUR family: *okage-sama de*. Because of you. It is meant to acknowledge our interdependence, and to deflect any honor or glory from ourselves to others. Because of you, I am what I am and I have this moment. Without you, it would not be possible.

We arrived in Yoshitomi-cho by train and were greeted by relatives we had never met. We had not expected them to come. We had planned on a taxi. But there they were, instantly recognizable—the same wide faces, the same bones—come to welcome the son of their deceased brother, who had gone to Hawaii so many years ago. We all stood on the train platform, eyes brimming with tears, bowing across the years and oceans, nobody touching, just bowing lower and bowing again, and again.

Later, in a little Kyushu village surrounded by luminous rice fields, my Japanese-American husband prayed at the shrine of his ancestors. Then, accompanied by a gaggle of nieces and nephews, we walked up the hill above the house to the graves where our name was inscribed in the ancient kanji characters in the stone. We stood for long moments. Yes, because of you. Okage-sama de.

MY STORY OF KYUSHU IS NOT JUST THE CHRONICLE OF OUR FAMILY, BUT THE story of Japan itself. Kyushu is the source of all that we think of as Japanese.

Recent archaeological evidence indicates that man has lived on

Kyushu, the most southern and western of Japan's four main islands, for about 12,000 years. The oldest pottery yet discovered was unearthed here. Legends say that when the sun goddess sent her grandson Ninigi to govern Japan, he touched down on Kyushu and wed the daughter of a local god. The date, as "calculated" by imperial historians, was February 11, 660 B.C. Ninigi's line continues to this day in the person of Emperor Akihito.

Ours is a much more modest family tree, although Jim remembers that his father, who owned a laundry business in Honolulu, was, from the minute he walked in the door of his house, accorded the deference due a god.

"Yes," Jim assured the houseful of relatives gathered on the tatami mats, "I grew up in a traditional Japanese home." They were greatly relieved, especially in view of me, his *hakujin*—Caucasian—wife.

Over a lunch prepared at great expense, including a bright red lobster on a swirled black-and-white plate, we fell easily into the universal patter of families. Aunt Tsuruye, cradling a grandson, recalled marrying into the family as a very young girl. "I was a picture bride to China. Noboru was living there during—excuse me, I'm very sorry—World War II. I didn't meet him until my wedding day. But I had seen a picture of him, and my family highly recommended him to me."

Noboru had passed away only two months ago, and clearly Aunt Tsuruye enjoyed speaking of him. We looked at photographs. Jim, the family announced, was the image of his father as they remembered him . . . Because of you, Kyushu.

WE HAD BOARDED THE BULLET TRAIN IN TOKYO, BOUND FOR KYUSHU WITH great expectations. We had been to Tokyo many times, and to paraphrase the 17th-century poet Basho, "Though in Japan, we longed for Japan." From the window, as the train slashed through a corridor of factories, smokestacks, power poles, and concrete condos that could have been steel mills (except for the laundry lines), it appeared that the Japanese had entirely devoured their land.

At the end of a long dark tunnel under the sea was Kyushu. We blazed into the light and almost wailed aloud at the sight: shipyards, fuel storage tanks, and the erector-set tangle of the city of Kitakyushu. Was there nothing left of serene green hills and lyrical streams tumbling over rounded rocks? Was nature in Japan reduced to a dish holding a bonsai garden?

But in time the factories began to thin out and the rice fields grew larger. There were hills in the hazy blue distance. We changed trains and boarded a lovely old coach with wooden floors, velvet seats, lacy curtains, and shiny brass fittings. The conductor wore a snappy plaid blazer. Ice cream was served, and pizza with shredded ginger. The train chugged along the coast, past fishermen in sturdy wooden boats. It puffed and tooted around curves and up into the woods and hills of Oita prefecture to our first stop, the town of Yufuin, where streets followed canals crossed by little bridges, and some of the older homes still had high thatched roofs.

Our inn was built of fragrant cedar. The entry stones and walkway through the bamboo had been washed for our arrival. Spare and cool, our cottage had mat floors and our own *furo*, a generous wooden tub that looked out on a wild, artfully untended garden. After a good soak at what seemed like the edge of the known world, we slipped into kimonos and geta, wooden clogs, and clumped clumsily off to dinner.

In the early morning we walked to a lake. Mist was rising slowly from the surface. The colors were palest pastel. The lotus blossoms had not yet stretched from their buds. An old torii gate inhabited the far shore. I had found Japan.

KYUSHU IS WHERE THE JAPANESE GO WHEN THEIR SOULS HAVE HAD ENOUGH OF Western ways. The hot springs resort area of Beppu attracts 12 million tourists a year, almost all of them Japanese. They come not only for the soothing thermal pools, but to relax and wear their *yukata*, fine cotton kimonos, in the street and walk in geta.

Every evening in the town of Hita, canopied barges festooned with pink paper lanterns floated out on the river. Those aboard wore kimonos and sat around on pillows, drinking sake and eating dinner. After more sake the singing started, and the barges rocked gently with dancing, the pink lanterns bobbing in the darkness as the happy music carried across the water.

It was summer, the season, it seemed, of traditional festivals. Throughout Kyushu people celebrated the O'bon, when Buddhists believe that departed ancestors return to earth, by dancing in the streets to the hypnotic sounds of flutes, drums, and samisen.

In the small town of Hondo we came upon a parade of floats, lanterns, troops of scouts, groups of women in traditional costumes with elaborate hats, and men dressed as fishermen and samurai. The parade had begun at sundown and would continue until sunup. We surrendered to exhaustion, returned to our inn, tumbled onto the plump futon sleeping mats, and fell asleep to the sound of distant gongs.

In Fukuoka, Kyushu's main city, we were among the 700,000 people who rose before dawn to cheer the race that culminates Hakata Gion Yamagasa, a two-week festival in July. Teams of men in loincloths and *happi* coats carried seven elaborate floats weighing a ton apiece through the streets in a contest so strenuous that no man could shoulder the burden for more than 50 seconds at a time. Each float demanded the effort of 600 men to run the three miles to the finish. With the battle cry "*Oissa! Oissa!*" the race started at precisely 4:59 A.M.

The shouts of the crowd became a brawny mantra as the men surged forward. "Oissa! Oissa!" The crowd roared in one voice. They threw buckets of water on the burdened men, whose bodies gleamed with sweat. It was a tradition that has been carried on for more than 750 years, little changed with the times. The floats used to be much bigger; now they have to fit beneath power lines.

All over Kyushu we continued to encounter that fabled soul of Japan, still alive and well, perhaps with a few modern compromises. At one hotel our kimonoed attendant, kneeling as she poured our tea, was interrupted by the bleep of her beeper tucked in her silk obi. "*Gomenasai,*" she whispered, "Excuse me." Then she raced for the telephone across the room.

At Yanagawa, the former home of a feudal lord has become a hotel with both Western and tatami rooms, and Western and Japanese gardens. The Japanese garden, laid out a century ago, is a miniature landscape of sea and mountains, meticulously manicured down to the smallest pine needle. The Western garden suffers some neglect. The gardeners do not understand it, or love it. Every evening the hotel staff sets out chairs and tables on the grass. They play old recordings of "Ramblin' Rose" and "Love Potion Number Nine," and serve nuts and beers to tourists from Taiwan.

In Yanagawa a friend who lives in a house with a Western-style living room and a Japanese-style dining room took us to the home of Kitahara Hakushu, a beloved Kyushu poet who died a half century ago. The house was set in a little garden. Shoji screens opened out to small views of immense delicacy. The rooms were almost devoid of furniture, the lighting soft, the wood fragrant with age. The floor timbers creaked beneath the tatami mats. The home had the sanctity and austerity of a monastery. Our friend said, "I come here often because I feel so Japanese when I'm here." From there we dashed off to pick up her young son, who had been playing in a baseball game that afternoon.

Change, whether the Japanese like it or not, is advancing at a two-steps-forward, one-step-backward pace. When we asked about a huge new home in the middle of the rice fields near Beppu, a young secretary complained, "Even Kyushu is beginning to change. In the last five years executives from Tokyo companies are coming and buying up the land and putting up big estates. It's driving our prices up."

But, she added, "Japanese society has not changed a lot since feudal times. We just have new shoguns."

The days of the original shoguns were days of glory for the privileged and saw a great flowering of Japanese arts. The era is celebrated annually in July at Kumamoto Castle, when lords in ancient armor, maids in pastel kimonos, saffron-robed monks, and brawny young men bearing golden shrines all come streaming out of the castle walls once again to parade about the city.

More than any part of the country, Kyushu seems to me to offer this sense of old Japan, not fossilized in temples and museums, but still lived and loved. One afternoon I watched a modern mother, impeccably suited in the latest fashion, gently instructing her teenage daughter in how to hold a cup of tea, turning the vessel so the best side faces the host.

The mother's gestures had the ease of a much practiced dance. The daughter was obedient though awkward, her movements jerky. But they would do this quietly together over the years until the girl grew into grace.

As a foreigner I sometimes got instruction of my own. In a small *ryokan* in Hondo, another kimonoed attendant, assuming I was unschooled in chopsticks, apologized profusely for not having any forks, and said she had no idea where she might get one.

The foreigner in Kyushu is treated with overwhelming kindness, rather like an indulged child. Our many breaches of esoteric etiquette are corrected and pardoned with a smile. The Japanese are well aware that their ways are confounding. The streets of some old cities were deliberately laid out to confuse an advancing army, and many of their customs seem to be founded on the same philosophy.

The simple act of bathing can be a perilous ritual fraught with potential for embarrassment. I was seduced into shedding both my inhibitions and my clothing by the reputation of Kyushu's thermal waters for curing an assortment of ailments and bestowing inner tranquility. We were in Beppu, the Reno of the Japanese hot springs towns, staying at the gargantuan Suginoi hotel, where the dinner buffet is as long as a football field and the *onsen* bath areas measured by the acre. I assumed that in a crowd my bath gaffes and my naked alien body might go unnoticed.

The Suginoi's two largest onsen are the Flower Bath and the Dream Bath. One for men and one for women, on alternating days. Even though I had received elaborate instructions regarding manners, I got off on the wrong foot by wearing my hall slippers too far into the dressing room. A woman, naked except for a shower cap, bowing and smiling, showed me, bowing and apologizing, where to leave them. I disrobed; then, with what I hoped was the dexterity of an origami artist, I folded my kimono neatly and placed it in a basket. Like the other women, I draped my oversize washcloth over strategic zones and entered the bath. It was an enormous steamy greenhouse with a high glass roof sheltering what appeared to be a slice of the Amazon rain forest. Several small pools were situated around a large pool, in the middle of the which, on a revolving dais, a great gilt Buddha and a Kannon, who represents compassion and mercy, presided over the bath, looking like a very big mama with a child on her knee.

It's amazing how our survival instincts work. In a fraction of a second I sized up the situation and, following the example of other new arrivals, walked to the side of the room, sat on a little stool, and poured a bucket of warm water over my body. I then soaped myself lavishly in an exercise that seemed to be part fastidiousness and part performance. Then I sat there, fully lathered, unable to figure out how to turn on the high-tech faucets with their multiple dials, hoses, and instructions printed in kanji characters. I turned a few dials. Nothing. Panic. Just when I feared I would have to retreat to my room trailing billows of suds, I spotted a shower with two easy faucets, rinsed, and stepped into the warm, soothing mineral bath. Buddha and Kannon looked down on me impassively as they made their rounds. The other women smiled, nodded, tried to engage me in conversation, then fell to talking quietly among themselves, lapsing into contented silences.

There is something very soothing about women bathing together, the soft chatter, the quietness, the absence of all maleness. Perhaps we harbor a lingering genetic memory of bathing together in rivers long ago. There was almost a solemnity, certainly a purity. I found that for hours afterward I was very alert, but my body seemed to be breathing on its own, as if I were asleep. I became an onsen junkie.

Some of Kyushu's springs are too hot to tame. They spit out of the ground, hissing over rocks in steamy rivers. At Ibusuki even the beach is thermally heated; people come from all over the country for sand baths.

Volcanic activity is at the very heart of Kyushu. One volcanic mountain, Aso, near Kumamoto city, has the largest caldera in the world. Within its crater are several towns and villages, a rail line, and lush rolling farmland. Aso is also a national park. The cliffs of the crater fall away in green palisades amid waterfalls, crystal clear streams, and forests of cedar and bamboo.

Still, the natural world has competition. One evening from a grassy knoll we watched a high-tech sound-and-light show featuring a metal grid assaulted by colored lights and space music full of growls and squeaky voices. Afterward we promised Masako Tanaka, who had accompanied us to Aso, that we would awaken her before sunrise for a real show.

In the first light of day the hills were flooded in gold and tinged with delicate pink. Wildflowers cheered the meadows—golden lilies, Queen Anne's lace, and gentians. Masako had never before gotten up to greet the dawn. She was ecstatic; "I didn't know it could be like this." She climbed a hill and stood with the sun beaming on her and lilies at her feet and threw her hands into the air, unable to contain her joy. Then she was overcome with bashfulness. She clutched her purse to her breasts in her typical shy posture, scarcely able to believe she had been so brash.

ANOTHER ACTIVE VOLCANO, SAKURAJIMA, SITS MENACINGLY IN THE MIDDLE OF a bay, belching ominous clouds of black smoke at the city of Kagoshima. In 1914 well water grew hotter for three days before Sakurajima blew. A wave of red-hot lava swept to the sea burying 600 homes and an ancient temple.

Nearly 8,000 people actually live on the volcano. Its ashes fall on the city like a dark and sooty snow, insidiously invading everything. You can feel and taste its grit. The streets must be hosed and swept daily. The children wear helmets to school in case of a rock rain. Homes, businesses, and tourist stops have shelter caves. The vegetation looks gray and burdened. Even the sun fails to cheer it.

One morning before dawn we looked across at the volcano from Castle Hill in Kagoshima. As light crept into the sky, people began to gather. Obviously they knew each other. Yet before they greeted their friends, they went to the edge of the hill and bowed to Sakurajima. A mournful gong rolled across the water announcing the rising sun.

At exactly 5:30 a man with a tape deck punched a button starting some inanely cheerful Looney Tunes music, and everyone began an aerobic exercise program that they all seemed to know well. And all the while Sakurajima mocked them, puffing out its smoke and ashes like an incorrigible old chain-smoker.

IN 1543 PORTUGUESE SAILORS, BLOWN OFF COURSE BY A TYPHOON, LANDED ON a small island off the coast of Kyushu. That chance contact brought other ships, and in 1549 tales of an alluring heathen land inspired the famous missionary, Francis Xavier, to set out for Japan. Mixing prayer and politics was standard practice for Jesuit missionaries and as they worked to convert the Kyushu daimyo, or lords, they also established Portuguese trading posts on the island, first at Hirako and Sakai, then at the small fishing village of Nagasaki.

The port quickly grew into one of the most prosperous cities in the country. The Portuguese were called *nambanjin,* southern barbarians. In spite of their uncomplimentary name and their reputation for not bathing, *namban* dress, food, religion, art, and objects became the rage throughout Japan.

Perceiving the new religion as a threat, the ruling shogun expelled all missionaries from Japan in 1587, though little was done to enforce his edict. Ten years later, however, six foreign and twenty Japanese Christians were crucified for their faith in Nagasaki. And though it seemed for a time that that was the end of the persecutions, eventually the practice of Christianity was forced underground. By 1638 some 40,000 Christians had been massacred. A year later the Portuguese were expelled from the country, Japan was closed to foreigners, Japanese citizens living or traveling abroad were forbidden to return home, and all trade (now run exclusively by the Dutch) was confined to a small island in Nagasaki Bay. For two centuries Nagasaki was the nation's only peephole to the outside world.

The religious sanctions were finally withdrawn in 1873, but even before that thousands of hidden Christians had emerged in the Nagasaki region. For generations they had practiced their faith in secret. Even though they no longer knew the meaning of the Latin, they recited prayers in their homes in front of Kannon statues that have come to be called Maria Kannon. These statues were very Buddhist-looking but were venerated as Mary with the infant Jesus.

In the 20th century death again came to Kyushu by the thousands, this time for residents of all faiths. On August 9, 1945, three days after the first atom bomb destroyed Hiroshima, a second was detonated over Nagasaki, killing more than 70,000 people and erasing half of the city.

At Nagasaki's International Cultural Hall, which is really the atom bomb museum, we were served iced tea as film footage of the devastation unreeled. It was a landscape of horror shaped by heat that reached 7,200°F and a blast that roared through the city at the speed of sound. Factories became twisted metal, schools instant graveyards.

One single image from the footage stayed with me; not the relentless human agony (God help us, but we have seen so many piles of mutilated corpses on television) but instead a sweep of land with nothing on it living, not a tree, not a blade of grass as far as the eye could see, to the seared and burned mountains. Nothing moved. A wind would have stirred only ashes.

I stopped taking notes. It was unbearable to watch. Asked for my reaction, I could only say, "The angels must have wept." As we did.

Today, a simple black marble slab in the Peace Park marks the epicenter of the bomb. It has no inscription.

What can be said?

LATER WE CLIMBED TO THE TOP OF MOUNT INASA AND LOOKED DOWN ON Nagasaki, rebuilt around its bowl of a harbor and straying up the mountainsides. Resurrected, it is easily one of the prettiest cities in Japan. One evening we went to not one but two lantern festivals, to listen to the grandmothers sing favorite folk songs and watch pretty girls dance with fans. It was hot, and the din of people, gongs, and loud monotonous music was intoxicating. We stuffed ourselves with *mochi*, rice paste wrapped in seaweed, and in the heat of the night we succumbed to luridly colored shaved ice cones, crunchy, sweet, cold. Strings of lanterns, each painted with a message and the name of a donor, festooned the trees. Beneath these necklaces of light were game booths and souvenir stalls laden with wind chimes. "Win a gold fish or a Donald Duck mask," shouted the hawkers late into the night. Everyone was in a kimono or a happi coat. Tired children were draped on the shoulders of their fathers.

In the end Kyushu was the Japan we had longed for, the Japan where people still string lanterns from temple trees and dance in the streets in summer, where they care deeply about how a cup of tea is held, and where they welcome the stranger with an overwhelming hospitality that has less to do with honor and more to do with simple kindness.

Okage-sama de. Because of you, we remember who we are and what paths we have traveled. [1991]

Yogyakarta: Java's Cultural Cauldron

JEFF GREENWALD

CHRISTIAN KHAENIN ISN'T the sort of man who can take a hint. The unassuming innkeeper moved to Java back in 1962, a 17-year-old in search of gainful employment on the high seas. He worked aboard ships for 16 years, persevering boldly as one boat after another sank from beneath him.

Khaenin shook his head. "The third time was really terrible. Just south of the Philippines, our ship was literally torn in half by a cyclone. There were few survivors." He squinted out the window into the fog.

"I spent 36 hours floating," he continued, "until I was finally rescued. They brought me back to Java to recover. Some weeks later, out of the hospital, I met an old man.

"You have three times sunk!" he told me. "If you sail again you will be killed!"

"And that, you see, is why I ran to the mountain."

But the mountain Khaenin chose for his personal refuge can hardly be called a snug harbor. His lodge rests in the small village of Kaliurang, 25 kilometers north of Yogyakarta, on the southern slopes of Mount Merapi. *Gunung* Merapi—fire mountain—is the second-highest volcano on Java, and one of the most active and destructive in the world.

"How can you stand it?" I asked Khaenin, startled by the figures in the tourist brochure. "This book says that Merapi has a major eruption about every two years. Don't you realize it's just a matter of months—*hours*, maybe—before this whole village is up to its ears in lava?"

"Yes, yes, yes . . . ." He nodded emphatically. "Merapi is certainly dangerous. We know that. But we are not helpless. Those of us living on the southern slopes know that Merapi will never bring us harm."

"How so?"

"We never forget to honor the guardian spirit that lives in the center of the volcano. He, in turn, respects us and allows us to remain here."

"Guardian spirit? But aren't you . . . ."

"That's right," he cuts in. "I am a Christian, as my name implies. But I'm also Indonesian. The sultan of Yogyakarta is a great Moslem, but once every year you will see him climbing Merapi—a difficult task!—with a sack full of offerings: flowers, fruits, silks, even cigarettes. He throws these into the crater, and the spirit invariably accepts our gifts.

"He is a pious Moslem, but our sultan believes in this spirit," said Khaenin. "And so do I."

The pursuit of a spirit at the center of things was what drew me to Java in the first place—specifically to the region surrounding Yogyakarta, a village-cum-city 28 kilometers north of the Indian Ocean and an equal distance south of Merapi's angry maw.

When I first read about Yogya, I felt that my vague curiosity about Indonesia had finally found a focus. Yogyakarta was the pulse of Java, the age-old capital that had never lost its eminence as the crucible of traditional culture. From the riverbed rhythms of gamelan orchestras to meticulous choreography, the highest arts of Java found their highest level here.

Yogya seemed, from the literature, to be the very soul of Java . . . the "Kyoto of Indonesia." It was hard to believe at first sight. Maybe I had expected something more poetic—a few gardens, at least, or a nice fountain. No such amenities were forthcoming. Just as the spirit deep in Merapi is shrouded by clouds of foul sulfur, so is the spirit of Java obscured somewhere behind the frantic avenues and equatorial glare of Yogyakarta. On an island teeming with spirits, it's a difficult one to pin down.

"WHY YOGYA?" I ASKED. "WHY DOES THE MAINSTREAM OF JAVANESE CULTURE flow from this apparently conventional city?"

Dr. Soeroso, who had been poised for a complex query, seemed momentarily flustered by my simplicity. He held his hands slightly apart, as if to clap them; they were deeply lined, somehow in contrast to his squarish, administrative features.

"Because there is nothing conventional *about* Yogyakarta," he said, bringing his palms together. "It has not evolved this way by chance."

Soeroso, a political scientist and Doctor of Philosophy who recently founded—in Yogya—the Institute of Javanology, warmed to my question.

"Seen historically, it makes perfect sense. After the Dutch colonized Central and Eastern Java back in the mid-1700s, the native culture concentrated around the last remaining strongholds of traditional power: the sultans. There was one here in Yogya, and another 60 kilometers northeast, in Solo. But the influence of those sultans continued to shrink, finally concentrating in the *kratons*—the palace courts—right at the heart of each city.

"Yogya and Solo flourished, and for 200 years the kraton in each city patronized the fine arts. And we don't speak of it much," he said, "but there had always been an edge of competition between the sultanates, with each one's style subtly different from, but of course vastly superior to, the others.

"Still," Soeroso continued, "there was one profound difference, and it spelled the end of Solo's sultanate. In 1945, during the war, Yogya fought for national independence . . . but Solo sided with the Dutch!" Soeroso laughed briefly, almost a bark. "It's a pity," he said.

After independence, some rapid changes were made. The Solo sultanate was unceremoniously dissolved, while Yogya's was as duly rewarded. Special Region Yogyakarta was named a self-governing territory, with the "governorship"—that is, the sultanate—a lifetime position. It was a good investment in loyalty. Yogya served as Indonesia's first capital, and has continued through the volatile years since to be a whip of nationalism. Meanwhile, the region's special status remains a magnet for artists, dancers and craftsmen.

Before leaving Dr. Soeroso, I complimented him on a wall decoration. It was a *wayang kulit*, one of the traditional leather shadow-puppets that are so popular throughout Indonesia. This one, made near Yogya, was especially beautiful—the flattened buffalo hide was expertly carved and painted with brilliant, almost psychedelic detail.

"That is a representation of Wisangeni" Soeroso said. "He's my favorite. To me, you see, he symbolizes the central spirit of the Javanese. He is brave. He is honest, self-confident and democratic—almost to a fault!"

"To a fault? How do you mean?"

"Well, you see, there are strict levels of address in the Javanese language. But Wisangeni will never use the high, formal address. He will only use the level for one's peers, whether he is speaking to a sultan or a slave. Yet in spite of these awful transgressions," said Soeroso, "Wisangeni is extremely powerful. He is, in fact, The Man Who Cannot Be Killed."

I thought about Wisangeni a few days later when preparing to visit the kraton. The wayang puppets in general are clearly behavior models for the average Javanese, but could Wisangeni be a specific metaphor for the sultan of Yogya himself? The public advocate; the one who speaks boldly; the man who never dies? In spite of his exalted rank, the sultan seems to be the source of much grassroots fellowship in Yogya. Invisible yet ever present, his charisma is the center around which Yogya's artisans orbit.

The kraton itself is a strange place. The inner compound, which the sultan still visits, is spotlessly maintained, and trafficked by the privileged courtiers who live within. These men of the inner court move somberly through the quiet sanctums, unmistakable in their traditional dress: batik skirts, high-necked shirts and *destars*—tightly wrapped batik skullcaps with an egglike knob protruding at the rear. On their backs the men wear sacred *kris*—wavy-bladed daggers imbued with magical powers.

Far more alive is the ramshackle outer kraton, the area beyond the sultan's court yet still within the compound's four kilometers of massive masonry walls. Here we find the Bird Market, a labyrinthine bazaar of feathers and squawks, and the crumbling Water Palace, once an elaborate pleasure garden and now a catacombed ruin of algaed pools and crumbling stonework.

Thousands of artisans still make their homes in this sprawling outer kraton. They cater mostly to the tourist trade, selling the effluvia of Javanese arts and crafts, but it is possible to find more. Hemmed on all sides by the painfully pedestrian wares of their neighbors, a few top-notch batik artists flounder or flourish within the faded radiance of the palace court.

Batik is easily the best-known of the Indonesian arts. One draws or paints upon fabric, using hot wax as the medium. The fabric is then dyed, the hardened wax acting as a resist. By boiling off the wax and applying a second design, the pattern can be developed further; it is not uncommon for a modern batik artist to dye and rewax dozens of times. It's an involved and labor-intensive process, but the results can be breathtaking.

Batik clothing has been an essential part of the Javanese wardrobe since the 16th century. Recently, though, the technique has become an artistic end in itself. Purely ornamental "batik painting" has been done on Java since the 1950s, and the movement has clearly centered in Yogya.

I'd been forewarned that Yogyakarta would be a labyrinth of studios, galleries and even batik "museums," run by an uneven gamut of artists—but I had a string to follow. A friend in Nepal had recommended the work of a young artist named Tulus Warsito. I'd seen one of his paintings, and it had impressed me: a batik portrait of Ronald Reagan, oddly cartoonlike yet somehow astonishingly accurate.

"But you'll especially like his floating jellybean paintings," my friend had insisted.

The idea of traditional court technique put to use as a New-Wave art medium intrigued me. I arrived at Tulus's studio early and found the artist drinking tea in his gallery. He greeted me in excellent English. I liked him immediately, and I liked his work as well. The walls were full of paintings, an eclectic collage of still lifes, village scenes and landscapes, but my eye was instantly drawn to those uncanny abstracts previously described. Brightly colored, beanlike shapes seemed to balloon above the canvas, casting lifelike "shadows" below. I'd certainly seen nothing like these in the world of batik before.

"Batik paintings always looked very flat to me," Tulus explained. "I felt I needed to find something different. Conceptually, these works were inspired by the wayang puppets and the way their shadows are deliberately projected onto a screen. And I was also inspired, of course, by the American plastic arts."

Tulus, at 33, has had more direct experience with Western art than most of his elders, let alone his peers. After winning a University of Oregon art competition in 1975, his work was featured at dozens of exhibitions throughout the States, and in 1977 Tulus visited America to teach the art of batik at Yale's Pierson College. His work is highly evolved and technically excellent, but Yogya's response to his abstract style has been cool.

"I've tried to set up a dialogue," Tulus said, "but it's very hard to organize artists. They're trying to sell, not develop. They think I'm crazy, taking too many risks. But I think it's riskier to attack that glutted common market than to oppose it with something new."

Not far from the Tulus Gallery are the showrooms of Mr. K. Kuswadji. With his roundish face and penetrating regard, Kuswadji bears a slight resemblance to Picasso. Over 70, he is perhaps the oldest batik-painter in Yogya, and he makes the claim, in red ink on his resume, of being "the first Painter using Batik as the Medium of Painting." His subject matter is linked to Tulus's in that they both draw inspiration from wayang. But while Tulus plays with the basic idea of a shadow, Kuswadji paints the wayang characters themselves—exaggerated figures of gods, lovers and heroes, glowing with rich color.

Unlike Tulus, Kuswadji had few words of complaint. I could see why: The gallery walls were empty, riddled with bare nails. A busload of tourists had recently stopped by, and when the dust settled only one painting remained unsold.

"Yogya will always be the center of Javanese art," claimed Kuswadji. "And the kraton must always be active! Its spirit is always with the people of Yogya." Kuswadji has a tight connection with the palace. Between 1935 and 1960 he lived within the kraton's inner circle, a renowned classical dancer who simultaneously perfected the "rhythms of painting."

"Yogya is getting better as it gets older," Kuswadji claimed. "Creativity is on the increase, because the senior artists spur on and encourage the young ones. There have been good times in Yogya," he said, "but this time, right now, is the golden age for young artists." He smiled at the empty walls. "And for old artists, too."

LATE ONE NIGHT, AS THE STREETS EMPTIED AND THE SIDEWALK VENDORS packed their wares, I wandered on foot toward my hotel. Not far beyond the kraton, a group of *becak* (bicycle ricksha) drivers sat laughing and shouting on an eerily lit veranda. The source of the strange light, I discovered, was a television set; the drivers were watching an American cop show called "CHiPs." Despite the program's incomprehensible plot and utterly alien tongue, the mob cheered with manic enthusiasm at every change of scene.

At an earlier moment the light would have struck me as absurd. That evening, though, I had attended my first wayang kulit show, and I thought that I understood.

"Wayang" means "shadow," and the wayang kulit are flat, carved leather puppets that perform behind a blacklit screen. Their projected images loom and dissolve, and the audience is held spellbound by tales drawn from 3,000 years of mythology. The art itself has developed over two millennia, and even in modern Yogya there are few men more honored than the skillful *dalang*, or puppet master. As he manipulates the heroes, villains and clowns—providing plot, narration, dialogue, action and even special effects—the dalang assumes a god-like role. And he must have comparable physical endurance, for each full play takes at least six hours to perform.

The show I attended was in full swing when I left, and would continue in high style until dawn. Sponsored by Yogyakarta State Radio, it was held in an auditorium within the kraton walls, a cavernous, meanly lit hall full of clove cigarette smoke and the ceaseless, hypnotic rhythms of a huge gamelan orchestra. A babbling stream of people filtered in and out; this was obviously as much a social event as a cultural one.

One thing in particular surprised me. Although I'd expected most people to be watching the ghostly shadows, the screen was arranged so that the laboring dalang and his cadre of meticulously crafted puppets were revealed to the audience. It was only in a small area behind the screen that people would watch the action as it was meant to be seen. The majority of Yogyanese seemed to enjoy the dalang's ballet of movements more than the shadow show itself, rather like preferring *The Making of Star Wars* to the movie itself.

It was later that very same evening, then, that I came upon the assembly of becak drivers riveted to the television set. I paused for a moment and tried to watch the bizarre high jinks of the police drama through their eyes. Doing so, I experienced a deep and sympathetic awe: how outrageous these wayang clowns seem, and how mysterious the dalang!

Much is made of the puppet drama in Java, and with good reason. It is, as Bill Dalton points out in his *Indonesia Handbook*, "one of the strongest cultural traits to have survived through the recorded history of Indonesia." Nevertheless, my interest in puppets could only go so far. I developed instead a strong personal preference for the *wayang orang*—Yogyakarta's renowned dance tradition.

There is a staggering variety of dance performances offered in Yogya, and the Tourist Information Center provides a complimentary schedule. But the most ambitious program I saw was not on the weekly playbill at all. This was the seasonal Prambanan *Ramayana*, a four-night enactment of Asia's most beloved epic. Held every full moon of the dry season (May until October) just east of the city, the *Ramayana* is one of Yogya's most famous events, with the thousand-year-old temples of the Prambanan Plain serving as the backdrop.

I attended three of the four August episodes, piloting a senile rented motorcycle over the 17 kilometers of potholes leading from Yogya to Prambanan. It was an extravaganza, all right, complete with floodlights and a capacity crowd. Hundreds of dancers and two full gamelan orchestras held the stage. Entire grade schools had been drafted into Thespian service, a few deft strokes of the makeup brush turning the Class of '93 into armies of monkeys or demons. I sat low in the stands, thrilled by the dancing but unable to hear a thing, for the audience—composed largely of family and friends—erupted with catcalls and howls that did not abate till curtain. Banana skins and other organic matter sailed through the air, briefly catching the spotlight before flopping to the stage. I was not disappointed; this was lively entertainment. But I knew that what I was witnessing was not exactly the zenith of Yogyanese dance.

During the week that followed, I did a bit of exploring around Yogya and found examples of wayang orang on considerably more intimate stages. But the most exquisite dance I saw in Yogyakarta was performed in a small, back-alley theater called the Dalem Pujokusuman. These thrice-weekly performances are also based on the *Ramayana* epic, but what a difference from the Prambanan show! No more than five dancers occupy the stage at one time, and the scenes are abridged into compact, flawless vignettes.

I arrived early and slipped into the dressing room. Evening prayers bleated futilely from the nearby mosque as the performers donned their Hindu costumes. Never before had I witnessed such complete transformation. The young men had arrived in a flurry of laughter and backslapping, but departed for the stage a troupe of icicle-eyed immortals.

How to describe the dances and do them justice? They were, I decided, like moonlight taking human form. The men were crossbows of energy, taut yet somehow fluid; the women so exquisite they were painful to behold. I could hardly bear to gaze at their lovely, impassive faces, imagining on those blank canvases the colorings of passion. Sitting practically onstage and nursing a ginger tea (that tasted, like everything else in Yogya, faintly but unmistakably of chicken), I found myself entranced by something more than human. Every aspect of these dancers, from the tilt of their heads to the posture of their fingers, was letter-perfect.

After the performance I was invited to interview Mr. Rama Sasminta Maridawa, the company's director. He spoke no English, but one of his pupils, a statuesque woman from Louisiana, agreed to interpret.

My first question involved something that had been puzzling me all week. Throughout Asia, I'd found, young people were drawing away from traditional values to pursue the golden apples of the West. But the situation seemed to be reversed here in Yogyakarta. There were new movements in batik; the shadow plays were maintaining their appeal; and wayang orang seemed to be enjoying an historic vitality. Could Maridawa explain this phenomenon?

"It's true," he said. "More students are studying now than ever before. That is partly because the political situation is finally stable. But you must also remember this: the arts you speak of are not for performance's sake as much as for knowledge and spiritual growth. That still means a lot here."

"Our style of dancing," Maridawa continued, "is especially valuable, for it embodies all the basic principles of Javanese tradition. That is, how to channel one's energy in a more constructive way. And many foreigners come here too, which even further increases the pride of our local students."

"Why here?" I asked. "Why Yogya in particular?"

"Because they see something here—in the art and dance of Yogyakarta—that is lacking in their own countries."

"And what might that be?"

"The spirit," Maridawa replied without a moment's hesitation. "The spirit is here." [1986]

Contributors

RITA ARIYOSHI is an award-winning writer and the author of the Hawaii best-seller *Maui on My Mind.* She has always lived on an island.

BILL BARICH is the author of two books of nonfiction, *Laughing in the Hills* and *Traveling Light*, and a book of stories, *Hard To Be Good.*

CHRISTOPHER BUCKLEY is editor of *Forbes FYI* and most recently, the author of a novel, *Wet Work.* He lives in Washington, D.C., with his wife and daughter.

TIM CAHILL lives in Montana at some distance from the nearest island. He is the author of four books: *Buried Dreams, Jaguars Ripped My Flesh, A Wolverine Is Eating My Leg,* and *Road Fever.*

BRYAN DI SALVATORE lives in Montana.

WILLIAM ECENBARGER was a reporter, first for United Press International and later for the Philadelphia *Inquirer,* before becoming a free-lance writer in 1981. His articles have appeared in ISLANDS, *Reader's Digest, Esquire,* the *Washington Post,* and dozens of other magazines and newspapers. He lives in Cornwall, Pennsylvania.

FRANCES FITZGERALD is a frequent contributor to *The New Yorker* magazine. Her books include *Fire in the Lake: The Vietnamese and the Americans in Vietnam* and *Cities on a Hill.*

BARRY GIFFORD is the author of the Sailor and Lula novels, which include *Wild at Heart* and *Sailor's Holiday*. His other books include the novels *Port Tropique* and *Landscape With Traveler*; the short story collection *New Mysteries of Paris*; and a fictional memoir, *A Good Man To Know*. He lives in the San Francisco Bay Area.

HERBERT GOLD, who lives in San Francisco, has written frequently about Haiti and other French-speaking islands. His newest book is *Best Nightmare on Earth: A Life in Haiti*.

JEFF GREENWALD lives in Oakland, California. His most recent book, *Shopping for Buddhas*, is an account of the author's frantic search for inner peace. His work on Asian travel, culture, and political issues has appeared in a variety of national publications, and he is currently writing a book about Tibet's human ecology.

KIM HEACOX, a former park ranger and biologist, has lived in Alaska since 1979. His articles and photographs have appeared in dozens of magazines, and he has twice won the Lowell Thomas Award for excellence in travel journalism, the second time, in 1990, for his ISLANDS story on Alaska's ABC Islands.

CARL HOFFMAN, who lives in Washington, D.C., spent childhood weekends sailing on the Chesapeake Bay. He has written for ISLANDS, *Smithsonian*, *National Geographic Traveler*, and the *Washington Post*, among other publications.

PICO IYER is an essayist for *Time* and a frequent visitor to Cuba. His books include *Video Night in Kathmandu* and *The Lady and the Monk*.

JESSICA MAXWELL lives with her husband, four cats, and, at last count, seven bunnies, in a log cabin on a rural northwest island, because it looks more like Ireland than anywhere else in America.

CAROL McCABE is book editor of the *Providence Journal-Bulletin*. Her travel articles have appeared in ISLANDS, *National Geographic Traveler*, *Travel & Leisure*, the *New York Times*, and the *Los Angeles Times*, among others. She lives in Bristol, Rhode Island, on Narragansett Bay.

JOHN McKINNEY, an award-winning nature writer and *Los Angeles Times* columnist, is the author of six travel books, including his latest, *Walk Los Angeles: Adventures on the Urban Edge*.

LAWRENCE MILLMAN is the author of seven books, including *Last Places*, an account of his adventures in the North. An ethnologist as well as a travel writer, he is currently working on a collection of Labrador Indian tales.

JAN MORRIS has been writing books for 35 years, including works of history and travel, as well as two volumes of autobiography, and a novel. She lives in Wales.

CHARLES NICHOLL, whose writing has taken him from Colombia (*The Fruit Palace*) to Thailand (*Borderlines*), was born in London and now lives in Herefordshire with his wife and four children. Having recently completed his fifth book, *The Reckoning*, an investigation of the murder of the Elizabethan playwright and poet Christopher Marlowe, he is off to Ethiopia for his next one.

ADAM NICOLSON, who lives in London, is a prize-winning travel writer and journalist. He's currently working on a novel set in sixth-century Britain and planning a voyage in the Celtic Sea.

MICHAEL PARFIT is the author of *Chasing the Glory*, about a flight around the American landscape, and *South Light; A Journey to the Last Continent*, a nonfiction work about life in Antarctica. He has traveled five times to Antarctica and has flown his own aircraft to both the Amazon and the arctic. He has written often for ISLANDS magazine and writes regularly for *Smithsonian*, on topics as diverse as Tony Hillerman and parachuting.

BOB PAYNE, has written for *Conde Nast Traveler* and *Outside* and is a contributing editor at *Sail*. He specializes in subjects that will keep him as far as possible from his home in Boston during the winter.

PAMELA SANDERS lives on a farm just outside Newport, Rhode Island. Her satirical novel, *The Oriental Secretary*, is set in Moscow and will be published next year. She returned to Leningrad last winter to research another work of fiction.

JEFF SPURRIER is a writer living in Los Angeles who spends about one third of the year in Mexico, miles away from any good fishing.

PAUL THEROUX's best-selling travel books include *The Great Railway Bazaar, The Old Patagonian Express, The Kingdom by the Sea*, and *Riding the Iron Rooster*. His most recent work is the novel *Chicago Loop*.

ISLANDS Magazine is published bimonthly by Islands Publishing Co., 3886 State St., Santa Barbara, CA, 93105, (805) 682-7177.